Interventional Management of Migraines and Other Headache Disorders

Interventional Management of Migraines and Other Headache Disorders

Edited by

Ivan Urits

Southcoast Physician Group, Wareham, MA, United States

Jamal Hasoon

UTHealth McGovern Medical School,
Department of Anesthesiology, Critical Care and Pain Medicine,
Houston, TX, United States

ELSEVIER

ACADEMIC PRESS

An imprint of Elsevier

Academic Press is an imprint of Elsevier
125 London Wall, London EC2Y 5AS, United Kingdom
525 B Street, Suite 1650, San Diego, CA 92101, United States
50 Hampshire Street, 5th Floor, Cambridge, MA 02139, United States

ISBN: 978-0-443-23557-3

For information on all Academic Press publications visit our website at https://www.elsevier.com/books-and-journals

Publisher: Stacy Masucci
Acquisitions Editor: Tracey Lange
Editorial Project Manager: Billie Jean Fernandez
Production Project Manager: Vijayaraj Purushothaman
Cover Designer: Vicky Pearson Esser

Typeset by TNQ Technologies

Contents

Contributors

Amit Aggarwal
Department of Anesthesiology and Perioperative Medicine, University of Texas Medical Branch, Galveston, TX, United States

Sait Ashina
Department of Neurology, Beth Israel Deaconess Medical Center, Harvard Medical School, Boston, MA, United States; Comprehensive Headache Center, Department of Neurology, Beth Israel Deaconess Medical Center, Harvard Medical School, Boston, MA, United States; Department of Anesthesiology, Critical Care, and Pain Medicine, Harvard Medical School, Beth Israel Deaconess Medical Center, Boston, MA, United States

Anthony J. Batri
New York-Presbyterian Hospital, the University Hospitals of Columbia and Cornell, Columbia University Irving Medical Center, New York, NY, United States

Spencer Brodsky
Montefiore Medical Center, Albert Einstein College of Medicine, Bronx, NY, United States

Casey Brown
Department of Physical Medicine & Rehabilitation, University of Pittsburgh Medical Center, Pittsburgh, PA, United States

Ahish Chitneni
Department of Rehabilitation and Regenerative Medicine, New York-Presbyterian Hospital - Columbia and Cornell, New York, NY, United States; LSU Health Shreveport (Formerly LSU Health Sciences Center Shreveport), Department of Anesthesiology, Shreveport, LA, United States

Zachary Danssaert
Weill Cornell Department of Rehabilitation Medicine, New York, NY, United States; Columbia University Department of Rehabilitation and Regenerative Medicine, New York, NY, United States

Moises Dominguez
Department of Neurology, Weill Cornell Medical College, New York Presbyterian Hospital, New York, NY, United States

Alexandra Fonseca
Department of Anesthesiology, Perioperative and Pain Medicine, Harvard Medical School, Brigham and Women's Hospital, Boston, MA, United States

Shane Fuentes
Department of Rehabilitation Medicine, Mayo Clinic School of Graduate Medical Education, Rochester, MN, United States

Joe Ghorayeb
University of Medicine and Health Sciences, New York, NY, United States

Hannah W. Haddad
Louisiana State University Health Science Center, Louisiana State University Physical Medicine and Rehabilitation Residency, New Orleans, LA, United States

Jamal J. Hasoon
University of Texas Health Science Center, Department of Anesthesiology, Critical Care, and Pain Medicine, Houston, TX, United States; UTHealth McGovern Medical School, Department of Anesthesiology, Critical Care and Pain Medicine, Houston, TX, United States

Mihir Jani
Montefiore Medical Center Department of Rehabilitation, Bronx, NY, United States

Crystal Joseph
Beth Israel Deaconess Medical Center, Boston, MA, United States

Ricky Ju
Burke Hospital Department of Rehabilitation, White Plains, NY, United States

Sohyun Kang
Department of Rehabilitation Medicine, Columbia University Medical Center, Weill Cornell Medical College, NewYork-Presbyterian Hospital, New York, NY, United States

Alan David Kaye
Department of Anesthesiology, Louisiana State University Health Sciences Center, Shreveport, LA, United States; LSU Health Shreveport (Formerly LSU Health Sciences Center Shreveport), Department of Anesthesiology, Shreveport, LA, United States

Joshua S. Kim
Weill Cornell Medical College, New York, NY, United States

Richard W. Kim
Columbia University Department of Rehabilitation and Regenerative Medicine, Weill Cornell Department of Rehabilitation Medicine, New York, NY, United States; Department of Physical Medicine & Rehabilitation, NYP Colombia Cornell, New York, NY, United States

Crystal Li
University of Maryland, School of Medicine, Baltimore, MD, United States

Aila Malik
Department of Physical Medicine & Rehabilitation, the University of Texas Health Science Center, McGovern Medical School, Houston, TX, United States

Danielle N. Nguyen
St. Elizabeth's Medical Center, Boston, MA, United States

QueenDenise Okeke
Philadelphia College of Osteopathic Medicine, Philadelphia, PA, United States

Vwaire Orhurhu
University of Pittsburgh Medical Center, Susquehanna, Williamsport, PA, United States; MVM Health, East Stroudsburg, PA, United States

Edward Pingenot
LSU Health Shreveport (Formerly LSU Health Sciences Center Shreveport), Department of Anesthesiology, Shreveport, LA, United States

Anish Rana
University of Rochester Medical Center, Rochester, NY, United States

Christopher L. Robinson
Department of Anesthesiology, Perioperative, and Pain Medicine, Harvard Medical School, Brigham and Women's Hospital, Boston, MA, United States

Roshan Santhosh
Department of Physical Medicine & Rehabilitation, Larkin Community Hospital, South Miami, FL, United States; Larkin Community Hospital, South Miami, FL, United States

Thomas T. Simopoulos
Department of Anesthesiology, Critical Care, and Pain Medicine, Harvard Medical School, Beth Israel Deaconess Medical Center, Boston, MA, United States

Isaac Springer
Oklahoma State University Medical Center, Oklahoma State University Anesthesiology Residency, Tulsa, OK, United States

Philip M. Stephens
Department of Physical Medicine & Rehabilitation, University of Pittsburgh Medical Center, Pittsburgh, PA, United States

Melissa M. Sun
Penn State Health Milton S. Hershey Medical Center, Rowan Virtua School of Medicine, Hershey, PA, United States

Jeremy Tuttle
University of Rochester Medical Center, Rochester, NY, United States; Weill Cornell Medical College, New York, NY, United States

Ivan Urits
Southcoast Health, Brain and Spine, Wareham, MA, United States; Southcoast Physician Group, Pain Management, Wareham, MA, United States

Peter D. Vu
Department of Physical Medicine & Rehabilitation, the University of Texas Health Science Center, McGovern Medical School, Houston, TX, United States

Daniel Wang
MedStar Georgetown University Hospital, Medstar Health Internal Medicine Residency, Washington, DC, United States

Cyrus Yazdi
Department of Anesthesiology, Critical Care, and Pain Medicine, Harvard Medical School, Beth Israel Deaconess Medical Center, Boston, MA, United States

Tsan-Chen Yeh
California University of Science and Medicine, Colton, CA, United States; Riverside Community Hospital, Riverside, CA, United States

Sandeep Yerra
Montefiore Health Systems, Bronx, NY, United States

Changho Yi
McAllen Medical Center, University of Texas Rio Grande Valley Family Medicine Residency, Brownsville, TX, United States

Emily X. Zhang
Massachusetts General Hospital, Department of Anesthesiology, Critical Care, and Pain Medicine, Boston, MA, United States

Chapter | One

Introduction to migraine: Current concepts, definitions, and diagnosis

Moises Dominguez[1,*], Sait Ashina[2,3,4], Cyrus Yazdi[2], Thomas T. Simopoulos[2], Jamal J. Hasoon[5], Alan David Kaye[6], Christopher L. Robinson[7]

[1]Department of Neurology, Weill Cornell Medical College, New York Presbyterian Hospital, New York, NY, United States; [2]Department of Anesthesiology, Critical Care, and Pain Medicine, Harvard Medical School, Beth Israel Deaconess Medical Center, Boston, MA, United States; [3]Department of Neurology, Beth Israel Deaconess Medical Center, Harvard Medical School, Boston, MA, United States; [4]Comprehensive Headache Center, Department of Neurology, Beth Israel Deaconess Medical Center, Harvard Medical School, Boston, MA, United States; [5]UTHealth McGovern Medical School, Department of Anesthesiology, Critical Care and Pain Medicine, Houston, TX, United States; [6]Department of Anesthesiology, Louisiana State University Health Sciences Center, Shreveport, LA, United States; [7]Department of Anesthesiology, Perioperative, and Pain Medicine, Harvard Medical School, Brigham and Women's Hospital, Boston, MA, United States

INTRODUCTION

Headache disorders are a highly prevalent neurologic condition in the general population, affecting approximately 90% of people during their lifetime [1]. Headache is a common reason to seek medical attention in the primary care setting, and in the United States, it is the fourth leading cause of emergency department visits [2]. Headache disorders can have many causes and are categorized into primary and secondary [3]. Primary headache disorders are not attributed to an underlying medical condition, while secondary headache disorders are. Secondary causes of

*Contribution of Moises Dominguez will be in public domain.

Interventional Management of Migraines and Other Headache Disorders. https://doi.org/10.1016/B978-0-443-23557-3.00001-2

headache include head trauma, meningitis, intracerebral hemorrhage, and brain masses amongst many others. Medication overuse headache is an additional example of a secondary headache disorder [4]. According to the International Classification of Headache Disorders, third edition (ICHD-3), there are four primary headache disorders classified: migraine, tension-type headache, trigeminal autonomic cephalalgias (e.g., cluster headache), and other primary headache disorders (e.g., new daily persistent headache) [5].

Migraine is a chronic, common, highly disabling primary headache disorder with additional features [6,7]. Migraine treatment can take many forms and include lifestyle modification, behavioral therapies (e.g., biofeedback), pharmacologic (e.g., acute and preventive treatment), procedural (e.g., onabotulinumtoxinA and pericranial peripheral nerve blocks), and the use of neuromodulation devices. In recent years, interventional treatments for migraine have experienced a significant rise in popularity and utilization. These treatments can serve as both primary and supplementary approaches in the acute, transitional, and preventive management of migraine.

This book, therefore, aims to provide an overview of interventional management approaches for migraine, focusing on interventional techniques that add to the armamentarium of migraine treatment. These interventions encompass various minimally invasive procedures, neurostimulation techniques, and nonpharmacological approaches targeting the underlying mechanisms of migraine, aiming to provide long-term relief and improve patient outcomes and quality of life.

DEFINITION AND CLINICAL FEATURES

Migraine is a chronic, recurring, primary headache disorder usually characterized by disabling headache episodes with associated symptoms. According to the ICHD-3, migraine is characterized by headache attacks lasting 4–72 h when untreated or unsuccessfully treated, accompanied by at least two of the following symptoms: unilateral location, pulsating quality, moderate to severe pain intensity, and aggravation or avoidance of routine physical activity [5]. Additionally, during the headache attack, individuals must experience at least one of the following: nausea and/or vomiting, photophobia, or phonophobia. It is important to note that the headache cannot be attributed to any other underlying cause, confirming its diagnosis as a primary headache disorder [5].

Migraine can be divided into four phases: prodrome (or premonitory), aura, headache phase, and postdrome [8]. The migraine prodrome is experienced in almost 80% of patients and usually appears 24–48 h before headache onset [9,10]. Prodromal symptoms include fatigue, yawning, neck pain or stiffness, nausea, food cravings, anorexia, mental slowness and impaired concentration, and diarrhea [8]. The migraine aura describes one or more focal neurological deficits that are transient and reversible. Approximately one-quarter of patients experience an aura, with visual aura being the most common [11]. Migraine auras include visual, sensory, language, motor, brainstem, and retinal symptoms. A migraine aura usually lasts 5–60 min, and two or more auras can occur in succession. Visual aura is the most common aura and is usually described as a small visual disturbance that gradually expands over 5–60 min, with the margins likely having geometric shapes or zigzagging lines. Although it was traditionally thought that the migraine aura occurs before the headache, it has been recognized to also occur during or after a headache phase [12]. Additionally, patients can also experience a migraine with aura but without the headache phase [5,13]. The migraine headache phase usually lasts 4–72 h if untreated or unsuccessfully treated and often is unilateral and has a throbbing or pulsating quality. After the headache phase, patients can have a migraine postdrome, lasting hours to days. Migraine postdrome symptoms include cognitive difficulties, asthenia, and somnolence [8]. Note that the interictal phase is a period in between migraine attacks where the individual is usually relatively symptom free [14].

Migraine can be further classified based on the frequency of headache days a patient experiences. A patient with fewer than 15 headache days per month is classified as having episodic migraine. However, if the patient experiences headaches on 15 or more days per month, for over 3 months, with at least eight of those days being migraines, they are diagnosed with chronic migraine [5]. Chronic migraine is a debilitating headache and subtype of migraine. In 2010, onabotulinumtoxinA injections received FDA approval as a treatment for this condition.

Epidemiology

Globally, migraine is recognized as the second leading cause of disability and is responsible for 45.1 million years lived with disability [15,16]. Particularly in individuals under 50 years of age, migraine is the leading cause of disability worldwide, especially in women [16]. The American Migraine Prevalence and Prevention (AMPP) Study, a population-based study, revealed that the median age of migraine onset in women was 23.2 and 25.5 years in men, and migraine burden peaks between 35 and 39 years of age [16,17]. As a result, migraine affects people during the most productive years and can significantly impact their lives, including the workplace, relationships, and parenting [7].

It is estimated that chronic migraine has a prevalence of 1.4%–2.2% on a global scale [18]. The AMPP study has demonstrated that approximately 2.5% of individuals with episodic migraine transformed to chronic migraine in the subsequent year [19]. According to the Chronic Migraine Epidemiology and Outcomes (CaMEO) study, individuals with chronic migraine display a greater disability than episodic migraine [20]. Risk factors associated with progression from episodic to chronic migraine include a high headache day frequency, depression, frequent acute headache medication overuse, and cutaneous allodynia [21].

Pathophysiology of migraine

It is no longer believed that migraine is caused by intracranial dilatation of cerebral blood vessels, as previously suggested by the vascular theory of migraine. This theory also held that constriction of these blood vessels was responsible for the accompanying migraine aura. However, current understanding has moved away from this belief [8,11]. Activation of the trigeminal vascular system (TVS) plays a role in migraine pathophysiology.

The TVS consists of nociceptive neurons that originate from the trigeminal ganglion (TG) and upper cervical dorsal roots and innervate intracranial vascular structures, such as the large cerebral and pial vessels, and dura mater (Fig. 1.1). The anterior structures of the head receive sensory innervation primarily from the ophthalmic division (V1) of the trigeminal nerve, while the upper cervical roots mainly innervate the posterior structures [22]. These sensory structures converge trigeminal cervical complex (TCC), which comprises the trigeminal nucleus caudalis (TNC) and the dorsal horn of the upper cervical cord (C1−C3) from where second-order neurons project to various nuclei in areas such as the thalamus (mainly the ventroposterior medial nucleus), hypothalamus, brainstem, and basal ganglia [23−26].These nuclei then project to various cortical structures, such as the sensory, visual, auditory, insular, and parietal cortices. This, in turn, can account for associated symptoms of migraine, such as photophobia, phonophobia, nausea, and osmophobia [26,27].

NEEDS FIGURE FOR MIGRAINE PATHOPHYSIOLOGY

Peripheral stimulation of these nociceptive neurons results in the release of vasoactive neuropeptides, such as calcitonin gene-related peptide (CGRP), pituitary adenylate cyclase-activating polypeptide-38 (PACAP-38), and substance P causing activation of the TVS [28,29]. The release of these neuropeptides may be associated with a sterile inflammatory response, also known as

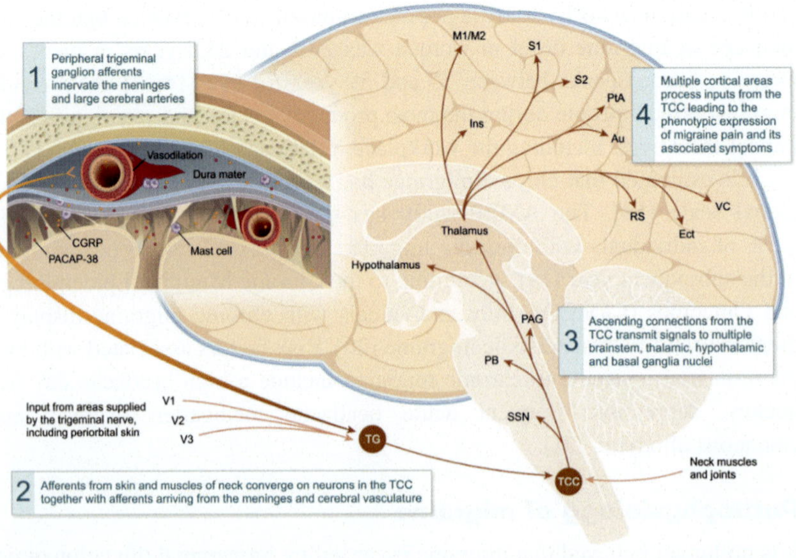

FIGURE 1.1 Visual description of the pathophysiology of migraine. A Phase-by-Phase Review of Migraine Pathophysiology - Dodick - 2018 - Headache: The Journal of Head and Face Pain - Wiley Online Library.

neurogenic inflammation, which may result in sensitization, a process by which neurons are hypersensitive to stimulation [30]. Peripheral sensitization is characterized by increased sensitivity of peripheral neurons in the TVS to dural stimulation, resulting from a lowered response threshold and heightened magnitude of responses to stimuli [31]. Clinically, peripheral sensitization may be responsible for the throbbing quality of migraine headaches and the exacerbation of head pain associated with increased intracranial pressure, such as coughing or bending over [31]. Central sensitization is characterized by hypersensitivity of second-order neurons in the TCC and related neurons in the central nervous system, such as thalamic nuclei. Central sensitization is thought to be responsible for development of allodynia during and outside of migraine attacks [32,33]. Cutaneous allodynia has been shown to be a risk factor for transformation of episodic migraine to chronic migraine [34].

Cortical spreading depression (CSD) is a phenomenon characterized by a self-propagating wave of depolarization that affects both neuronal and glial cells in the cerebral cortex, which may result in the headache and aura associated with migraine (Fig. 1.1) [35,36]. This depolarization wave, associated with a wave of hyperemia, spreads slowly across the cortex at 2–6 mm per minute, followed by inhibition of cortical activity, which is associated with oligemia. CSD is associated with the efflux of potassium, the influx of sodium and calcium, and the release of molecules such as glutamate and adenosine triphosphate [31,37]. There is evidence to support the notion that CSD activates trigeminal afferents, which leads to headache in migraine [38] (Fig. 1.2).

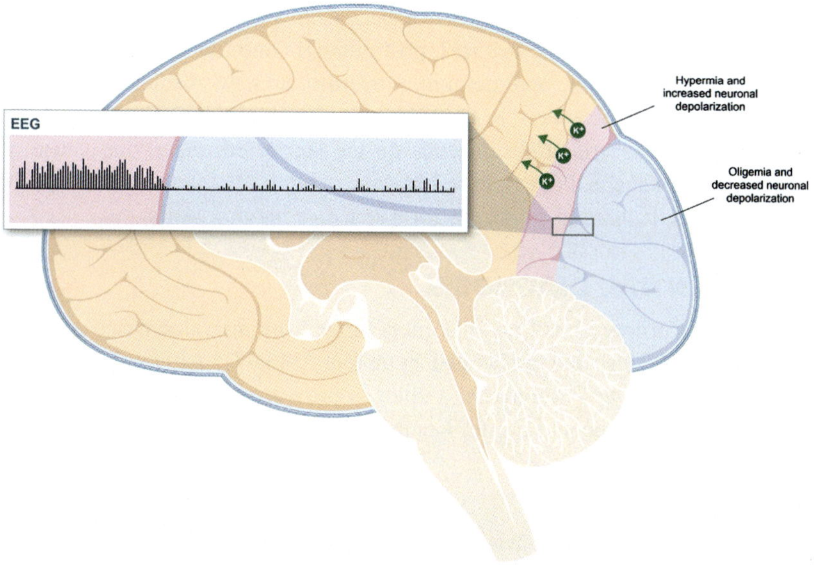

FIGURE 1.2 Visual representation of cortical spreading depression. A Phase-by-Phase Review of Migraine Pathophysiology - Dodick - 2018 - Headache: The Journal of Head and Face Pain - Wiley Online Library.

CGRP is believed to play an important role in the development of migraine [26]. This neuropeptide, consisting of 37 amino acids, is found in the sensory nerves (A-delta and C fibers) that innervate the blood vessels of the meninges. It has been associated with enhancing glutamatergic signaling and has been observed to be present at elevated plasma levels in the jugular vein during migraine attacks [39,40]. Evidence suggests that the administration of CGRP can induce migraine headache, specifically in individuals with a history of migraine [41,42]. CGRP is believed to be involved in transmitting pain signals from the intracranial blood vessels to the central nervous system through the trigeminovascular pathway. Additionally, it plays a role in the vasodilatory aspect of neurogenic inflammation [43]. Given CGRPs critical role in the pathophysiology of migraine, both acute and rescue CGRP targeted treatments have been developed to reduce the burden of migraine.

Diagnosis

Currently, there is no specific diagnostic test that identifies individuals with migraine. Therefore, the diagnosis is clinical and depends upon a compatible medical history, neurologic examination, and meeting diagnostic criteria specified by the International ICHD-3 (Table 1.1) [5,44].

For most of migraine patients, neuroimaging is typically unnecessary. However, it is crucial to remain vigilant for any concerning signs or symptoms requiring further diagnostic evaluation. The mnemonic SNNOOP10 can be a helpful tool for identifying potential red flags [45]. The presence of **s**ystemic symptoms (including fever), history of **n**eoplasm, **n**eurological deficit or dysfunction, sudden **o**nset headache, **o**lder age (after 50 years), **p**attern change, **p**ositionality, **p**apilledema, **p**rogressive headache, **p**regnancy or **p**uerperium, **p**ainful eye with autonomic features, **p**osttraumatic onset headache, **p**athology of the immune system, and **p**ainkiller overuse or new medication, should prompt healthcare providers to consider appropriate diagnostic studies [45]. The type of diagnostic study depends on the suspected underlying cause of the patient's headache. For example, obtaining a brain MRI with contrast would be appropriate if there is a concern for a space-occupying lesion.

DIFFERENTIAL DIAGNOSIS

Briefly, other primary headache disorders differ from migraine based on location, characteristics, and associated symptoms [46,47]. Although the main focus of this chapter is migraine, a chart is provided to differentiate it from tension-type headache and cluster headache.

CONCLUSION

Migraine is a prevalent and disabling chronic neurological disorder with a complex pathophysiology. Migraine is more than just a headache, involving notable sensory disturbances and distinct phases. These symptoms can severely impact

Table 1.1 Provides diagnostic criteria for both migraine with and without aura, as per the ICHD-3.

ICHD-3 criteria for migraine without aura [5]

1. At least five attacks fulfilling criteria B through D
2. Headache attacks lasting 4–72 h (untreated or unsuccessfully treated)
3. Headache has at least two of the following characteristics:
   a. Unilateral location
   b. Pulsating quality
   c. Moderate or severe pain intensity
   d. Aggravation by or causing avoidance of routine physical activity (e.g., walking or climbing stairs)
4. During headache, at least one of the following:
   a. Nausea, vomiting, or both
   b. Photophobia and phonophobia
5. Not better accounted for by another ICHD-3 diagnosis

ICHD-3 criteria for migraine with aura [5]

1. At least two attacks fulfilling criterion B and C
2. One or more of the following fully reversible aura symptoms:
   a. Visual
   b. Sensory
   c. Speech and/or language
   d. Motor
   e. Brainstem
   f. Retinal
3. At least three of the following six characteristics:
   a. At least one aura symptom spreads gradually over $\geq$5 min
   b. Two or more symptoms occur in succession
   c. Each individual aura symptom lasts 5–60 min
   d. At least one aura symptom is unilateral
   e. At least one aura symptom is positive
   f. The aura is accompanied, or followed within 60 min, by headache
4. Not better accounted for by another ICHD-3 diagnosis

[redacted]

Location	Usually unilateral	Bilateral	Unilateral
Pain quality	Pulsating, throbbing	Pressure or tightening	"Stabbing," severe
Headache duration	4–72 h	30 min to 7 days	15 min to 3 h
Associated symptoms	Nausea, vomiting, photophobia, phonophobia. May have aura	None	Ipsilateral lacrimation, conjunctival injection, rhinorrhea, nasal congestion, miosis, restlessness, or agitation.

an individual's quality of life. Diagnosis of migraine is primarily based on clinical evaluation. The subsequent chapters in this book will explore the interventional management of migraine.

REFERENCES

[1] B.K. Rasmussen, R. Jensen, M. Schroll, J. Olesen, Epidemiology of headache in a general population—a prevalence study, J. Clin. Epidemiol. 44 (1991) 1147−1157.

[2] R.C. Burch, S. Loder, E. Loder, T.A. Smitherman, The prevalence and burden of migraine and severe headache in the United States: ustatistics from government Health surveillance studies, Headache J. Head Face Pain 55 (2015) 21−34.

[3] P. Rizzoli, W.J. Mullally, Headache, Am. J. Med. 131 (2018) 17−24.

[4] S. Ashina, et al., Medication overuse headache, Nat. Rev. Dis. Prim. 9 (2023) 5.

[5] Headache classification committee of the international headache society (IHS) the international classification of headache disorders, 3rd edition, Cephalalgia 38 (2018) 1−211.

[6] R.C. Burch, D.C. Buse, R.B. Lipton, Migraine: epidemiology, burden, and comorbidity, Neurol. Clin. 37 (2019) 631−649.

[7] M. Ashina, et al., Migraine: epidemiology and systems of care, Lancet 397 (2021) 1485−1495.

[8] M.D. Ferrari, et al., Migraine, Nat. Rev. Dis. Prim. 8 (2022) 2.

[9] L. Kelman, The premonitory symptoms (prodrome): a tertiary care study of 893 migraineurs, Headache J. Head Face Pain 44 (2004) 865−872.

[10] K. Laurell, et al., Premonitory symptoms in migraine: a cross-sectional study in 2714 persons, Cephalalgia 36 (2015) 951−959.

[11] M. Ashina, Migraine, N. Engl. J. Med. 383 (2020) 1866−1876.

[12] J.M. Hansen, et al., Migraine headache is present in the aura phase, Neurology 79 (2012) 2044.

[13] D.R. Shah, S. Dilwali, D.I. Friedman, Current aura without headache, Curr. Pain Headache Rep. 22 (2018) 77.

[14] K.-P. Peng, A. May, Redefining migraine phases − a suggestion based on clinical, physiological, and functional imaging evidence, Cephalalgia 40 (2020) 866−870.

[15] T. Vos, et al., Global, regional, and national incidence, prevalence, and years lived with disability for 328 diseases and injuries for 195 countries, 1990−2016: a systematic analysis for the Global Burden of Disease Study 2016, Lancet 390 (2017) 1211−1259.

[16] L.J. Stovner, et al., Global, regional, and national burden of migraine and tension-type headache, 1990−2016: a systematic analysis for the Global Burden of Disease Study 2016, Lancet Neurol. 17 (2018) 954−976.

[17] W.F. Stewart, C. Wood, M.L. Reed, J. Roy, R.B. Lipton, Cumulative lifetime migraine incidence in women and men, Cephalalgia 28 (2008) 1170−1178.

[18] J.L. Natoli, et al., Global prevalence of chronic migraine: a systematic review, Cephalalgia 30 (2009) 599−609.

[19] M.E. Bigal, et al., Acute migraine medications and evolution from episodic to chronic migraine: a longitudinal population-based study, Headache J. Head Face Pain 48 (2008) 1157−1168.

[20] R.B. Lipton, A. Manack Adams, D.C. Buse, K.M. Fanning, M.L.A. Reed, Comparison of the chronic migraine epidemiology and outcomes (CaMEO) study and American migraine prevalence and prevention (AMPP) study: demographics and headache-related disability, Headache J. Head Face Pain 56 (2016) 1280−1289.

[21] R.C. Burch, D.C. Buse, R.B. Lipton, Migraine, Neurol. Clin. 37 (2019) 631−649.

[22] M.A.R. Arbab, L. Wiklund, N.A. Svendgaard, Origin and distribution of cerebral vascular innervation from superior cervical, trigeminal and spinal ganglia investigated with retrograde and anterograde WGA-HRP tracing in the rat, Neuroscience 19 (1986) 695−708.

[23] T. Bartsch, P.J. Goadsby, Stimulation of the greater occipital nerve induces increased central excitability of dural afferent input, Brain 125 (2002) 1496−1509.

[24] T. Bartsch, P.J. Goadsby, Increased responses in trigeminocervical nociceptive neurons to cervical input after stimulation of the dura mater, Brain 126 (2003) 1801–1813.
[25] T. Bartsch, P.J. Goadsby, The trigeminocervical complex and migraine: current concepts and synthesis, Curr. Pain Headache Rep. 7 (2003) 371–376.
[26] D.W. Dodick, A phase-by-phase review of migraine pathophysiology, Headache J. Head Face Pain 58 (2018) 4–16.
[27] R. Noseda, M. Jakubowski, V. Kainz, D. Borsook, R. Burstein, Cortical projections of functionally identified thalamic trigeminovascular neurons: implications for migraine headache and its associated symptoms, J. Neurosci. 31 (2011) 14204.
[28] F.M. Amin, et al., Investigation of the pathophysiological mechanisms of migraine attacks induced by pituitary adenylate cyclase-activating polypeptide-38, Brain 137 (2014) 779–794.
[29] K. Messlinger, M.J.M. Fischer, J.K. Lennerz, Neuropeptide effects in the trigeminal system: pathophysiology and clinical relevance in migraine, Keio J. Med. 60 (2011) 82–89.
[30] R. Burstein, Deconstructing migraine headache into peripheral and central sensitization, Pain 89 (2001) 107–110.
[31] R. Burstein, R. Noseda, D. Borsook, Migraine: multiple processes, complex pathophysiology, J. Neurosci. 35 (2015) 6619.
[32] R. Burstein, D. Yarnitsky, I. Goor-Aryeh, B.J. Ransil, Z.H. Bajwa, An association between migraine and cutaneous allodynia, Ann. Neurol. 47 (2000) 614–624.
[33] S. Ashina, A. Melo-Carrillo, E. Szabo, D. Borsook, R. Burstein, Pre-treatment non-ictal cephalic allodynia identifies responders to prophylactic treatment of chronic and episodic migraine patients with galcanezumab: a prospective quantitative sensory testing study (NCT04271202), Cephalalgia 43 (2023) 03331024221147881.
[34] M.A. Louter, et al., Cutaneous allodynia as a predictor of migraine chronification, Brain 136 (2013) 3489–3496.
[35] A.A.P. Leao, Spreading depression of activity in the cerebral cortex, J. Neurophysiol. 7 (1944) 359–390.
[36] A. Charles, Advances in the basic and clinical science of migraine, Ann. Neurol. 65 (2009) 491–498.
[37] A. Charles, K.C. Brennan, Cortical spreading depression—new insights and persistent questions, Cephalalgia 29 (2009) 1115–1124.
[38] H. Bolay, et al., Intrinsic brain activity triggers trigeminal meningeal afferents in a migraine model, Nat Med 8 (2002) 136–142.
[39] P.J. Goadsby, L. Edvinsson, R. Ekman, Vasoactive peptide release in the extracerebral circulation of humans during migraine headache, Ann. Neurol. 28 (1990) 183–187.
[40] A.F. Russo, Calcitonin gene-related peptide (CGRP): a new target for migraine, Annu. Rev. Pharmacol. Toxicol. 55 (2015) 533–552.
[41] L.H. Lassen, et al., Cgrp may play A causative role in migraine, Cephalalgia 22 (2002) 54–61.
[42] J.M. Hansen, A.W. Hauge, J. Olesen, M. Ashina, Calcitonin gene-related peptide triggers migraine-like attacks in patients with migraine with aura, Cephalalgia 30 (2010) 1179–1186.
[43] A. Charles, P. Pozo-Rosich, Targeting calcitonin gene-related peptide: a new era in migraine therapy, Lancet 394 (2019) 1765–1774.
[44] A.K. Eigenbrodt, et al., Diagnosis and management of migraine in ten steps, Nat. Rev. Neurol. 17 (2021) 501–514.
[45] T.P. Do, et al., Red and orange flags for secondary headaches in clinical practice, Neurology 92 (2019) 134–144.
[46] A. May, et al., Cluster headache, Nat. Rev. Dis. Prim. 4 (2018) 18006.
[47] S. Ashina, et al., Tension-type headache, Nat. Rev. Dis. Prim. 7 (2021) 24.

Botulinum toxin injection for migraine and other headache disorders

Moises Dominguez[1,*], Sait Ashina[2,3], Cyrus Yazdi[2], Thomas T. Simopoulos[2], Jamal J. Hasoon[4], Ivan Urits[5], Alan David Kaye[6,7], Christopher L. Robinson[8]

[1]Department of Neurology, Weill Cornell Medical College, New York Presbyterian Hospital, New York, NY, United States; [2]Department of Anesthesiology, Critical Care, and Pain Medicine, Harvard Medical School, Beth Israel Deaconess Medical Center, Boston, MA, United States; [3]Comprehensive Headache Center, Department of Neurology, Beth Israel Deaconess Medical Center, Harvard Medical School, Boston, MA, United States; [4]UTHealth McGovern Medical School, Department of Anesthesiology, Critical Care and Pain Medicine, Houston, TX, United States; [5]Southcoast Physician Group, Pain Management, Wareham, MA, United States; [6]Department of Anesthesiology, Louisiana State University Health Sciences Center, Shreveport, LA, United States; [7]LSU Health Shreveport (Formerly LSU Health Sciences Center Shreveport), Department of Anesthesiology, Shreveport, LA, United States; [8]Department of Anesthesiology, Perioperative, and Pain Medicine, Harvard Medical School, Brigham and Women's Hospital, Boston, MA, United States

INTRODUCTION

Globally, headache is recognized as a highly disabling and prevalent medical condition [1]. Migraine, a chronic and debilitating headache disorder, affects over 1 billion people worldwide, irrespective of regional, cultural, or socioeconomic differences [2,3]. It is the second most common cause of disability worldwide and the leading cause of disability among young women [4,5]. Due to its widespread occurrence and significant impact on society, headaches are a common reason for patients to seek medical care [6,7].

*Contribution of Moises Dominguez will be in public domain.

Interventional Management of Migraines and Other Headache Disorders. https://doi.org/10.1016/B978-0-443-23557-3.00002-4

Migraine is a primary headache disorder characterized by recurring, moderate-to-severe pulsating or throbbing headaches associated with symptoms such as photophobia, phonophobia, and nausea and vomiting [8]. Migraine is classified as episodic when there are <15 headache days per month. On the other hand, chronic migraine (CM) occurs when there are ≥ 15 headache days per month, at least eight of those days meeting International Classification of Headache Disorders-3 (ICHD-3) diagnostic criteria for migraine, for at least 3 months [8]. Across the glove, CM has a prevalence of about 2% and, according to the American Migraine Prevalence and Prevention (AMPP) study, 2.5%–3% of episodic migraine (E.M.) transition to CM annually [9,10]. CM is more disabling than E.M. and is associated with increased use of healthcare resources and higher direct and indirect costs [11,12]. Despite the availability of numerous treatment options, patients may continue to experience inadequate relief from their migraine episodes, encounter adverse events associated with pharmacologic therapies, or face limited options due to underlying medical conditions or long-term medication use, potentially impacting the efficacy of certain oral migraine treatments [13].

Botulinum toxin, a potent neurotoxic protein, is produced by the gram-positive anaerobic bacteria of the genus Clostridium (*Clostridium botulinum, Clostridium butyrricum, Clostridium baratii,* and *Clostridium argentinensis*) [14]. Ingestion of this bacterial toxin can eventually result in diffuse flaccid paralysis and respiratory failure. There are seven major botulinum toxin serotypes, which are designated from A-G; however, only serotypes A and B are currently used in clinical practice. Initially utilized for the treatment of strabismus, botulinum toxin has since expanded its applications to include various conditions such as hyperhidrosis, cervical dystonia, blepharospasm, and muscle spasticity [15–17].

Of note, Dr. Binder, a facial plastic surgeon, observed that patients receiving cosmetic botulinum toxin also had an additional benefit for their migraine [18]. This notable finding eventually led to the pivotal Phase III Research Evaluating Migraine Prophylaxis Therapy (PREEMPT) study, which demonstrated the efficacy of botulinum toxin in the management of chronic migraine [19,20]. Subsequently, the application of botulinum toxin has been investigated in other primary headache disorders such as cluster headache and tension-type headache. In the present investigation, we provide a comprehensive review on botulinum toxin as a treatment for chronic migraine, and its potential use in other headache disorders.

MECHANISM OF ACTION

The botulinum toxin exhibits favorable pharmacologic properties, including neurospecificity, limited diffusion when injected locally at a targeted site, and reversibility of its effects. These features are notably seen with botulinum toxin A, which is considered a safe, efficacious injectable neuromodulator for treating hyperfunctioning nerve terminals leading to human disease [21].

Botulinum toxin A injection into a muscle prevents muscle contraction by inhibiting the release of acetylcholine from presynaptic neurons at the neuromuscular junction. Botulinum toxin A is a peptide that has one heavy and light chain. Initially, the botulinum toxin binds to peripheral nerve terminals through its heavy chain, which subsequently becomes endocytosed within an endocytic vesicle [21]. The protonation of botulinum toxin facilitates the translocation of its light chain into the neuronal cell's cytosol. This light chain possesses metalloprotease activity and, depending on the serotype, selectively cleaves specific components of soluble N-ethylmaleimide-sensitive factor attachment protein receptor (SNARE) in peripheral nerve terminals. These SNARE proteins encompass vesicle-associated membrane protein (VAMP; also known as synaptobrevin), synaptosome-associated protein of 25 kd (SNAP-25), and syntaxin. Botulinum toxins A and E specifically target SNAP-25, whereas botulinum toxins B, D, F, and G primarily cleave VAMP. On the other hand, botulinum toxin C can cleave both syntaxin and SNAP-25. Ultimately, cleaving SNARE proteins disrupts the essential function of the SNARE complex, impairing the fusion of synaptic vesicles with the neuronal cell membrane and preventing acetylcholine from being released into the neuromuscular junction. This process effectively prevents subsequent cascading events that result in muscle contraction.

Several botulinum toxin preparations are approved for clinical use, most of which are based on serotype A1 and one on serotype B1. OnabotulinumtoxinA, which is marketed as Botox by Allergan, an AbbVie company, is U.S. Food and Drug Administration (FDA) approved for the treatment of chronic migraine. OnabotulinumtoxinA cleaves SNAP-25, which allows for the stable formation, but nonfunctional SNARE complex, impairing exocytosis of certain neurotransmitters (e.g., acetylcholine) (Fig. 2.1). While the mechanism of action of

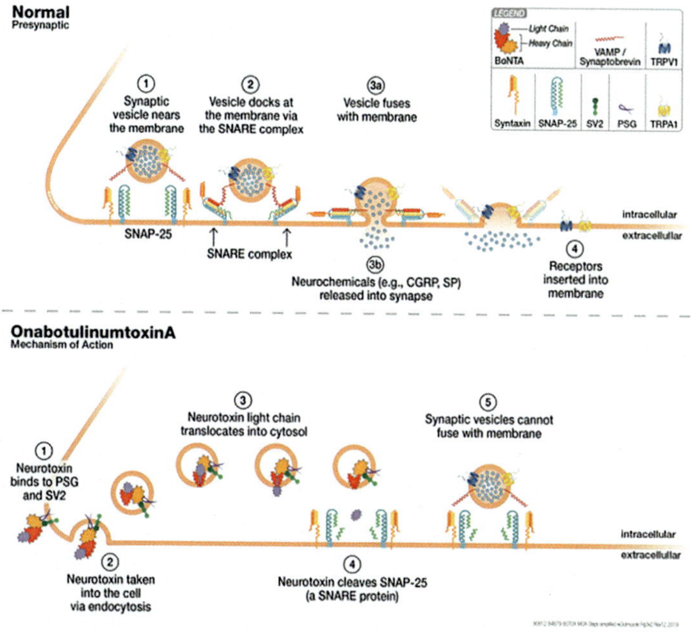

FIGURE 2.1 Mechanism of action of onabotulinum toxin A. https://www.ncbi.nlm.nih.gov/pmc/articles/PMC7496564/.

onabotulinumtoxinA is well understood at the level of the neuromuscular junction, there is still more to be learned on its impact on the sensory system related to migraine and other headache disorders. To understand how this toxin affects pain modulation, it is important to mention that in addition to small synaptic vesicles containing small molecules such as acetylcholine and glutamine in presynaptic terminals of peripheral nerves, there are also large dense core vesicles that contain proteins, such as calcitonin gene-related peptide (CGRP), pituitary adenylate cyclase activity, and receptors, such as transient receptor potential cation channel subfamily V and A member 1, that play a critical role in pain signaling [22,23]. OnabotulinumtoxinA has been shown to impair the release of certain pronociceptive molecules, such as calcitonin-gene-related peptides and substance P [24,25].

Pain associated with migraine begins peripherally, where there is activation of nociceptive neurons that innervate the dura mater, resulting in the release of vasoactive and proinflammatory neuropeptides and neurotransmitters (e.g., CGRP, PACAP-38, and nitric oxide) [26–29]. These molecules have been implicated in the pathophysiology of migraine. OnabotulinumtoxinA reduces the release of these neuropeptides and neurotransmitters at the synaptic cleft [24,30,31]. This raises the question of how extracranial injection of onabotulinumtoxinA affects intracranial neurons. This question continues to be investigated, but a possible explanation is that onabotulinumtoxinA affects the branches of trigeminal axons that cross intracranially to outside the cranium or from cervical axons that cross from outside the cranium to the inside [32–35]. Ultimately, onabotulinumtoxinA injection appears to reduce the activation of peripheral and central pathways involved in mediating migraine headaches through complex mechanisms, which is believed to be the result of decreased peripheral input rather than transsynaptic transfer [36].

OnabotulinumtoxinA is strategically injected into seven muscle groups containing fibers responsible for sensory information in the head and neck, which are supplied by trigeminal and cervical sensory neurons [37]. These muscle groups include corrugators, procerus, frontalis, temporalis, occipitalis, cervical paraspinal, and trapezius muscles. Upon intramuscular injection into 31–39 sites, the toxin targets sensory nerves within a specific region, which are the supraorbital, supratrochlear, auriculotemporal, zygomaticotemporal, greater and lesser occipital, third occipital, and supraclavicular nerves [37] (Fig. 2.2).

CLINICAL EVIDENCE

In 2005, two studies assessed the efficacy of botulinum toxin type A in the treatment of chronic daily headache (CDH), which is a heterogeneous group of primary headache disorders characterized by 15 or more days per month of headache [38,39]. One of these studies was a randomized, double-blind, placebo-controlled, parallel-group clinical study [38]. In this study, although patients in the treatment arm had an improved mean change from baseline in the number of headache-free days compared with the placebo arm, the improvement did not meet statistical significance at day 180. Nevertheless, there was a statistically significant improvement in secondary efficacy measures at day 180.

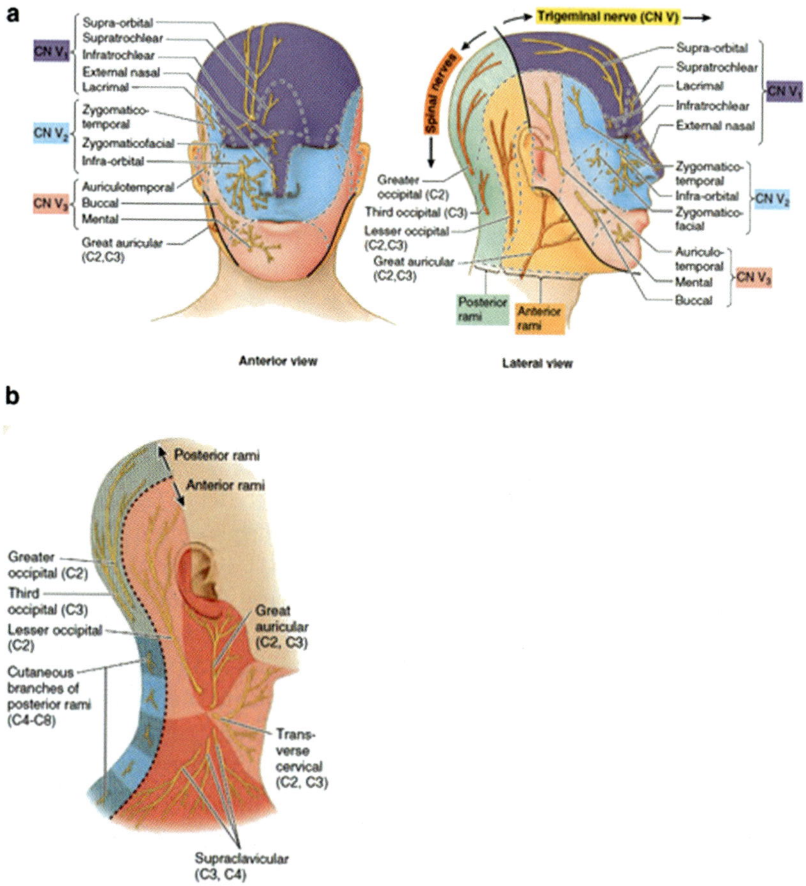

FIGURE 2.2 Sensory nerves and locations of the head and neck. https://headachejournal.onlinelibrary.wiley.com/doi/10.1111/head.13074.

These included a reduction of 50% or more in the frequency of headache days compared with the baseline over a 30-day period, and the mean change from baseline in headache frequency was greater in the treatment arm over the 30-day period [38]. The other study was also a randomized, double-blind, placebo-controlled, parallel-group clinical study where eligible patients received 225, 150, and 75 units of botulinum toxin type A in the treatment arm [39]. The primary endpoint of mean change from baseline in the number of headache-free days per 30-day period was unmet. For secondary efficacy endpoints, there were no statistically significant differences in the percentage of patients with a decrease of at least 50% in the number of headache days between the BoNTA treatment groups and placebo nonresponders at day 180.

In 2008, researchers conducted a study to examine how effective a fixed dose and site administration of 100 units of botulinum toxin type A could be for individuals with chronic migraine, specifically those without acute headache medication overuse [40]. Compared with placebo, botulinum toxin type A demonstrated a clinical benefit in individuals with chronic migraine. For example, patients in the botulinum toxin type A group showed a significant reduction in monthly migraine episodes, and twice as many patients in this group experienced a substantial decrease of 50% or more in their migraine episodes compared with those who received the placebo.

Two pivotal trials, PREEMPT 1 and PREEMPT 2, demonstrated evidence that onabotulinumtoxinA was effective for the treatment of chronic migraine [19,20]. PREEMPT 1 was a phase 3 study with two phases: a 24-week, double-blind, parallel-group, placebo-controlled phase and a 32-week open-label phase. Their primary outcome, which was a mean change from baseline in headache episode frequency at week 24, was not met. However, there was a statistically significant improvement in headache and migraine days, triptan use, change from baseline in moderate/severe headache days, total cumulative headache hours on headache days, and disability and functioning [19]. PREEMPT 2, also a phase 3 study and relatively larger than PREEMPT 1 (705 vs. 679 patients), demonstrated a statistically significant improvement in the primary outcome (changes from baseline in frequency of headache days) [20]. Analysis of combined data from PREEMPT 1 and PREEMPT 2 demonstrated favorable results for the onabotulinumtoxinA treatment group [41]. In patients treated with onabotulinumtoxinA, the frequency of headache days was reduced compared with baseline, and there were positive effects found in most secondary outcome measures except for the frequency of acute pain medication use. A meta-analysis additionally supported the effectiveness of OnabotulinumtoxinA injections as a preventive treatment for chronic migraine and subgroup analysis showed reversal of medication overuse [42,43]. It is worth noting that it can take up to three treatment cycles before a clinical benefit is noticed [44].

The effectiveness of onabotulinumtoxinA injection as a preventive treatment for chronic tension-type headaches remains unclear. A recent meta-analysis of 11 controlled trials supported its benefit in multiple measurement areas, including standardized headache intensity, headache frequency, daily headache duration, and the use of acute pain medication; however, the quality of evidence was rated as low-moderate [45]. A systemic literature review was conducted to determine botulinum toxin A's effectiveness in managing trigeminal neuralgia [46]. The review included four randomized, double-blind, placebo-controlled trials and concluded that botulinum toxin type A treatment is safe and effective; however, the quality of most of the trials may be poor [47]. Botulinum toxin A may be a viable treatment option for patients with refractory primary headaches other than CM who do not respond to standard therapies; nevertheless, the evidence is limited but it may have potential benefits in refractory tension-type headache, trigeminal autonomic cephalalgias, primary stabbing headache, nummular headache, hypnic headache, and new daily persistent headache after unsuccessful trials with other medications [48,49].

PATIENT SELECTION AND INJECTION TECHNIQUE

According to the ICHD-3 criteria, chronic migraine is defined as experiencing 15 or more headache days per month for over 3 months, with at least 8 days displaying migraine-like features. OnabotulinumtoxinA injections are usually considered when initial pharmacologic preventive treatments, such as antihypertensives, antidepressants, and antiseizure medications which are associated with side effects, are contraindicated, or fail to manage chronic migraine effectively. The European Headache Federation recommends using onabotulinumtoxinA to treat chronic migraine after an individual has tried two to three other preventive migraine treatments [50].

The PREEMPT injection and dose paradigm was based on two prior studies [38,39]. The PREEMPT protocol involves the administration of onabotulinumtoxinA at 31 specific intramuscular injection sites, targeting muscles such as the corrugators, procerus, frontalis, temporalis, occipitalis, cervical paraspinal, and trapezius (Fig. 2.3). These injections are performed every 12 weeks using a 30-

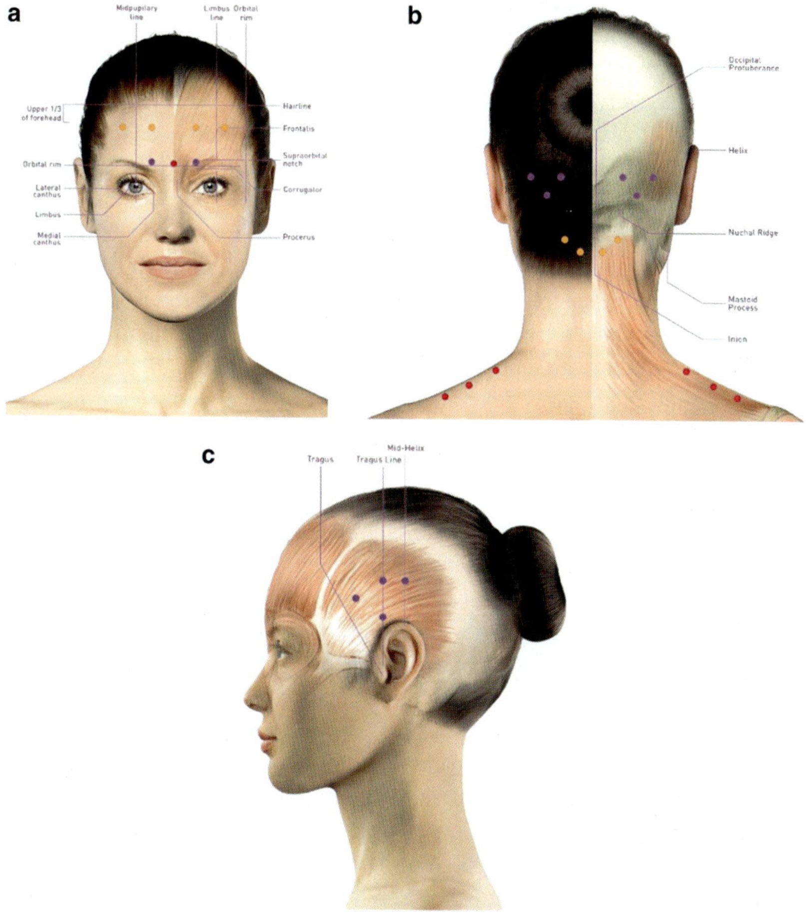

FIGURE 2.3 PREEMPT protocol injection sites. (A) The corrugator muscle is indicated by purple dots, the procerus muscle by the *red* dot, and the frontalis muscle by *orange* dots. (B) The occipitalis area is marked with *purple* dots, the cervical paraspinal area with orange dots, and the trapezius muscle with red dots. (C) Purple dots identify the temporalis muscle. Reproduced with permission from AbbVie Inc. from *Headache* published by Wiley Periodicals, Inc. on behalf of American Headache Society.

gauge, 0.5-inch needle, delivering 5 units (0.1 mL) of the onabotulinumtoxinA solution [19,20]. In PREEMPT 1 and PREEMPT 2, the remaining 40 units could be administered through a follow-the-pain strategy, specifically in the temporalis, occipitalis, and/or trapezius. Therefore, 155 units of onabotulinumtoxinA toxin are administered with the option of adding 40 units in a "follow-the-pain."

Adverse effects and safety profile

Onabotulinumtoxin A for chronic migraine is regarded as safe and efficacious. The most common adverse reaction is neck pain and headache following injection. In the pooled results from PREEMPT 1 and PREEMPT 2, adverse events were reported by 62.4% of patients treated with onabotulinumtoxinA injection compared with 52.7% of patients in the placebo group [41]. Most patients who experienced adverse events described them as mild to moderate in severity. Only a small percentage of patients discontinued the study, with 3.8% in the onabotulinumtoxinA group and 1.2% in the placebo group. The most common treatment-related adverse events include neck pain (6.7%), muscular weakness (5.5%), eyelid ptosis (3.3%), injection site pain (3.2%), and headache (2.9%).

In a recently published review, the safety of onabotulinumtoxinA in chronic migraine was investigated through a systematic review and meta-analysis of randomized clinical trials [51]. This study examined the treatment-related adverse events in randomized clinical studies of onabotulinumtoxinA compared with placebo and other treatments. OnabotulinumtoxinA produced relatively more treatment-related adverse events than placebo but fewer than oral topiramate. This finding further supported the notion that onabotulinumtoxinA injection for chronic migraine is generally safe and well-tolerated. However, there was high heterogeneity among the included studies (I [2] = 96%), likely due to significant variations in study sample sizes and compounded by high dropout rates.

Further research is needed to determine the safety of onabotulinumtoxinA injections for chronic migraine during pregnancy. Related to the current lack of evidence, it is common practice to discontinue onabotulinumtoxinA injections when planning for pregnancy or upon confirmation of pregnancy. A prospective study in a headache clinic followed 45 pregnant patients, collecting data on pregnancy outcomes. Among these patients, 32 chose to continue onabotulinumtoxinA treatment. Over 9 years, the study found that, with the exception of one miscarriage in the treatment group, all neonates were delivered at full term with normal birth weight and without any congenital malformations [52]. A 24-year retrospective review of the Allergan safety database indicates that fetal defects in mothers exposed to onabotulinumtoxinA before or during pregnancy (2.7%) are similar to the rates observed in the general population [53]. Monitoring of pregnancy outcomes in women exposed to onabotulinumtoxinA is ongoing. Regarding lactation, there is currently no available data on the use of onabotulinumtoxin injections for chronic migraine during breastfeeding. It is unclear what effect it would have on the infant being breastfed. However, information from LactMed indicates that onabotulinumtoxinA is not detectable systemically following intramuscular administration, which suggests that its presence in breast milk is unlikely [54,55]. Additional research is necessary to provide appropriate guidance for this specific patient population.

The safety and efficacy of onabotulinumtoxinA in the treatment of chronic migraine in pediatric patients requires further investigation and research. In the literature review conducted by Marcelo et al., the majority of the included studies indicated a relatively favorable safety profile of onabotulinumtoxinA [56].Two randomized controlled trials assessing the efficacy and safety of onabotulinumtoxinA injections in pediatric patients have suggested that the treatment is well-tolerated. Both studies reported no treatment-related serious adverse events, and the rates of adverse events were low [57,58].

Limited research exists regarding the safety of onabotulinumtoxinA in elderly individuals with chronic migraine. The Chronic Migraine OnabotulinuMtoxinA Prolonged Efficacy open Label (COMPEL) study is a long-term, open-label prospective study that collected data from multiple centers from various countries. This study, which involved patients aged 18-73, further added to the long-term safety and efficacy of onabotulinumtoxinA over a 2-year period, potentially extending to patients aged 65-73 [59]. In a post hoc analysis, real-life prospective data from 16 European headache centers were examined [60]. The study focused on the treatment of elderly patients (65 years of age and older) with chronic migraine using onabotulinumtoxinA for three cycles. The analysis revealed that there were no significant differences in discontinuation rates or adverse events between individuals aged 18-64 and those aged 65 and older.

LONG-TERM OUTCOMES AND TREATMENT OPTIMIZATION

A pooled analysis of the 56-week PREEMPT clinical program has yielded valuable insights into the long-term treatment efficacy and safety of onabotulinumtoxinA in chronic migraine patients [61]. The analysis demonstrated that repeat treatments of five or fewer cycles of onabotulinumtoxinA were considered effective, safe, and well tolerated, suggesting that the durability of benefit was noted beyond 1 year. A separate study assessed the sustained effectiveness of at least 1 year of onabotulinumtoxinA treatment in a real-world setting [62]. Participants were included if they met the diagnostic criteria of chronic migraine with or without analgesic overuse, insufficient response to treatment, and unable to tolerate or have contraindications to beta-blockers, topiramate, amitriptyline, flunarizine, and valproic acid. The study demonstrated that ~70% of participants with chronic migraine had a sustained response to onabotulinumtoxinA for years. They noted that 9.3% of patients lost their clinical response after 1 year. With the exception of two cases where local muscle atrophy occurred after treatment exceeding 5 years, the observed adverse events are consistent with those reported in short-term clinical trials.

In a retrospective study, patients with chronic migraine who had medication overuse or were concurrently using oral preventive treatment and received a minimum of five cycles of onabotulinumtoxinA injections following the PREEMPT protocol, were analyzed [63]. This study demonstrated that it is achievable to discontinue medication overuse and oral preventive treatment with long-term use of onabotulinumtoxinA injection. A recent systematic literature review revealed that onabotulinumtoxinA therapy had shown positive effects in real-world clinical studies, including reductions in the number and frequency of headache or migraine days, concomitant medication use, and improvements in HIT-6 scores across a diverse range of patients [64]. Another study reported that onabotulinumtoxinA treatment was well tolerated over 108 weeks of treatment [65].

Although patients benefit from regular onabotulinumtoxinA injections every 12 weeks, some patients experience worsening headaches leading up to their next scheduled injection. This observation is commonly framed as a "wear-off" phenomenon. In one retrospective chart review of 143 patients with chronic migraine treated with onabotulinumtoxinA over a 2-year period, wear-off was present in 62.9% of patients [66]. Wear-off usually occurs 2–4 weeks before their next injection [66]. A prospective report of 24 patients noted that wear-off typically occurred in the eighth week following injection [67]. The retrospective chart review indicated that there were fewer mean units of onabotulinumtoxinA injected administered in the wear-off group compared with the no wear-off group (166.0 ± 13.1 vs. 173.4 ± 10.3, $P = 0.0005$) [66]. When wear-off is present, adding the "follow the pain" protocol in addition to the PREEMPT protocol may be helpful in patients with chronic migraine who respond to onabotulinumtoxinA injection for less than 3 months [63]. During a 2-year treatment period in patients with chronic migraine and medication overuse headaches, higher doses of onabotulinumtoxinA (195 units) are superior to standard dosing (155 units) while maintaining a comparable safety and tolerability profile [68]. During this wear-off period, intramuscular ketorolac, craniocervical peripheral nerve blocks, and intravenous or oral corticosteroids have been used as a bridge therapy until their next injection [66].

FUTURE DIRECTIONS AND CONCLUSION

OnabotulinumtoxinA injection has emerged as a promising, effective treatment option for patients with chronic migraine, expanding the range of available treatments for migraine prevention. Its administration every 12 weeks contributes to its efficacy and promotes improved treatment adherence. The favorable side effect profile associated with onabotulinumtoxinA further enhanced its appeal to patients as a therapeutic option. Further research is needed to explore the efficacy and safety of onabotulinumtoxinA in episodic migraine in adults, chronic and episodic migraine in pediatric patients, and other headache and facial pain disorders. Moreover, it is vital to investigate the potential clinical efficacy and safety profiles of other forms of botulinum toxin, such as abobotulinumtoxinA, prabotulinumtoxinA, and incobotulinumtoxinA [69]. Furthermore, the safety of onabotulinumtoxinA during pregnancy and breastfeeding requires further investigation to ensure appropriate use in these specific populations.

REFERENCES

[1] T. Vos, S.S. Lim, C. Abbafati, K.M. Abbas, M. Abbasi, M. Abbasifard, et al., Global burden of 369 diseases and injuries in 204 countries and territories, 1990–2019: a systematic analysis for the global burden of disease study 2019, Lancet 396 (10258) (2020) 1204–1222, https://doi.org/10.1016/s0140-6736(20)30925-9.

[2] Vos, et al., The Lancet, 2016, p. 2016, r aly, y li, lo.

[3] A.H. Ropper, M. Ashina, Migraine, N. Engl. J. Med. 383 (19) (2020) 1866–1876, https://doi.org/10.1056/nejmra1915327.

[4] T.J. Steiner, L.J. Stovner, R. Jensen, D. Uluduz, Z. Katsarava, Migraine remains second among the world's causes of disability, and first among young women: findings from GBD2019, J. Headache Pain 21 (1) (2020), https://doi.org/10.1186/s10194-020-01208-0.

[5] M. Ashina, Z. Katsarava, T. Phu Do, D.C. Buse, P. Pozo-Rosich, A. Özge, A.V. Krymchantowski, E.R. Lebedeva, K. Ravishankar, S. Yu, S. Sacco, S. Ashina, S. Younis, T.J. Steiner, R.B. Lipton, Migraine: epidemiology and systems of care, Lancet 397 (10283) (2021) 1485–1495, https://doi.org/10.1016/s0140-6736(20)32160-7.

[6] Z. Katsarava, M. Mania, C. Lampl, J. Herberhold, T.J. Steiner, Poor medical care for people with migraine in Europe – evidence from the Eurolight study, J. Headache Pain 19 (1) (2018), https://doi.org/10.1186/s10194-018-0839-1.

[7] S. Bösner, S. Hartel, J. Diederich, E. Baum, Diagnosing headache in primary care: a qualitative study of GPs' approaches, Br. J. Gen. Pract. 64 (626) (2014) e532, https://doi.org/10.3399/bjgp14x681325.

[8] Headache classification committee of the international headache society (IHS) the international classification of headache disorders, 3rd edition, Cephalalgia 38 (1) (2018) 1–211, https://doi.org/10.1177/0333102417738202.

[9] J.L. Natoli, A. Manack, B. Dean, Q. Butler, C.C. Turkel, L. Stovner, R.B. Lipton, Global prevalence of chronic migraine: a systematic review, Cephalalgia 30 (5) (2010) 599–609, https://doi.org/10.1111/j.1468-2982.2009.01941.x.

[10] M.E. Bigal, D. Serrano, D. Buse, A. Scher, W.F. Stewart, R.B. Lipton, Acute migraine medications and evolution from episodic to chronic migraine: a longitudinal population-based study, Headache J. Head Face Pain 48 (8) (2008) 1157–1168, https://doi.org/10.1111/j.1526-4610.2008.01217.x.

[11] W.F. Stewart, G.C. Wood, A. Manack, S.F. Varon, D.C. Buse, R.B. Lipton, Employment and work impact of chronic migraine and episodic migraine, J. Occup. Environ. Med. 52 (1) (2010) 8–14, https://doi.org/10.1097/JOM.0b013e3181c1dc56.

[12] R.B. Lipton, A.M. Adams, D.C. Buse, K.M. Fanning, M.L. Reed, A comparison of the chronic migraine epidemiology and outcomes (CaMEO) study and American migraine prevalence and prevention (AMPP) study: demographics and headache-related disability, Headache J. Head Face Pain 56 (8) (2016) 1280–1289, https://doi.org/10.1111/head.12878.

[13] R.B. Lipton, R.A. Nicholson, M.L. Reed, A.B. Araujo, D.H. Jaffe, D.E. Faries, D.C. Buse, R.E. Shapiro, S. Ashina, M. Janelle Cambron-Mellott, J.C. Rowland, E.M. Pearlman, Diagnosis, consultation, treatment, and impact of migraine in the US: results of the OVERCOME (US) study, Headache J. Head Face Pain 62 (2) (2022) 122–140, https://doi.org/10.1111/head.14259.

[14] T.J. Smith, K.K. Hill, B.H. Raphael, Historical and current perspectives on Clostridium botulinum diversity, Res. Microbiol. 166 (4) (2015) 290–302, https://doi.org/10.1016/j.resmic.2014.09.007.
[15] M. Naumann, N.J. Lowe, C.R. Kumar, H. Hamm, Botulinum toxin type A is a safe and effective treatment for axillary hyperhidrosis over 16 months: a prospective study, Arch. Dermatol. 139 (6) (2003) 731–736, https://doi.org/10.1001/archderm.139.6.731.
[16] A.B. Scott, Botulinum toxin injection into extraocular muscles as an alternative to strabismus surgery, Ophthalmology. 87 (10) (1980) 1044–1049, https://doi.org/10.1016/S0161-6420(80)35127-0.
[17] D.M. Simpson, M. Hallett, E.J. Ashman, C.L. Comella, M.W. Green, G.S. Gronseth, M.J. Armstrong, D. Gloss, S. Potrebic, J. Jankovic, B.P. Karp, M. Naumann, Y.T. So, S.A. Yablon, Practice guideline update summary: botulinum neurotoxin for the treatment of blepharospasm, cervical dystonia, adult spasticity, and headache, Neurology 86 (19) (2016) 1818–1826, https://doi.org/10.1212/WNL.0000000000002560.
[18] W.J. Binder, A. Blitzer, M.F. Brin, Treatment of hyperfunctional lines of the face with botulinum toxin A, Dermatol. Surg. 24 (11) (1998) 1198–1205, https://doi.org/10.1111/j.1524-4725.1998.tb04098.x.
[19] S.K. Aurora, D.W. Dodick, C.C. Turkel, R.E. DeGryse, S.D. Silberstein, R.B. Lipton, H.C. Diener, M.F. Brin, OnabotulinumtoxinA for treatment of chronic migraine: results from the double-blind, randomized, placebo-controlled phase of the PREEMPT 1 trial, Cephalalgia 30 (7) (2010) 793–803, https://doi.org/10.1177/0333102410364676.
[20] H.C. Diener, D.W. Dodick, S.K. Aurora, C.C. Turkel, R.E. DeGryse, R.B. Lipton, S.D. Silberstein, M.F. Brin, OnabotulinumtoxinA for treatment of chronic migraine: results from the double-blind, randomized, placebo-controlled phase of the PREEMPT 2 trial, Cephalalgia 30 (7) (2010) 804–814, https://doi.org/10.1177/0333102410364677.
[21] M. Pirazzini, O. Rossetto, R. Eleopra, C. Montecucco, J.M. Witkin, Botulinum neurotoxins: biology, pharmacology, and toxicology, Pharmacol. Rev. 69 (2) (2017) 200–235, https://doi.org/10.1124/pr.116.012658.
[22] J. Meng, J. Wang, M. Steinhoff, J.O. Dolly, TNFα induces co-trafficking of TRPV1/TRPA1 in VAMP1-containing vesicles to the plasmalemma via Munc18-1/syntaxin1/SNAP-25 mediated fusion, Sci. Rep. 6 (2016), https://doi.org/10.1038/srep21226.
[23] I. Devesa, C. Ferrándiz-Huertas, S. Mathivanan, C. Wolf, R. Luján, J.-P. Changeux, A. Ferrer-Montiel, αCGRP is essential for algesic exocytotic mobilization of TRPV1 channels in peptidergic nociceptors, Proc. Natl. Acad. Sci. USA 111 (51) (2014) 18345–18350, https://doi.org/10.1073/pnas.1420252111.
[24] P.L. Durham, R. Cady, R. Cady, Regulation of calcitonin gene-related peptide secretion from trigeminal nerve cells by botulinum toxin type A: Implications for migraine therapy, Headache J. Head Face Pain 44 (1) (2004) 35–43, https://doi.org/10.1111/j.1526-4610.2004.04007.x.
[25] J. Purkiss, M. Welch, S. Doward, K. Foster, Capsaicin-stimulated release of substance P from cultured dorsal root ganglion neurons: involvement of two distinct mechanisms, Biochem. Pharmacol. 59 (11) (2000) 1403–1406, https://doi.org/10.1016/s0006-2952(00)00260-4.
[26] K. Messlinger, M.J.M. Fischer, J.K. Lennerz, Neuropeptide effects in the trigeminal system: pathophysiology and clinical relevance in migraine, Keio J. Med. 60 (3) (2011) 82–89, https://doi.org/10.2302/kjm.60.82Germany.
[27] F.M. Amin, A. Hougaard, H.W. Schytz, M.S. Asghar, E. Lundholm, A.I. Parvaiz, P.J.H. De Koning, M.R. Andersen, H.B.W. Larsson, J. Fahrenkrug, J. Olesen, M. Ashina, Investigation of the pathophysiological mechanisms of migraine attacks induced by pituitary adenylate cyclase-activating polypeptide-38, Brain 137 (3) (2014) 779–794, https://doi.org/10.1093/brain/awt369.
[28] K. Meßlinger, U. Hanesch, M. Baumgärtel, B. Trost, R.F. Schmidt, Innervation of the dura mater encephali of cat and rat: ultrastructure and calcitonin gene-related peptide-like and substance P-like immunoreactivity, Anat. Embryol. 188 (3) (1993) 219–237, https://doi.org/10.1007/BF00188214.

[29] D.W. Dodick, A phase-by-phase review of migraine pathophysiology, Headache J. Head Face Pain 58 (S1) (2018) 4–16, https://doi.org/10.1111/head.13300.

[30] M.J. Welch, J.R. Purkiss, K.A. Foster, Sensitivity of embryonic rat dorsal root ganglia neurons to Clostridium botulinum neurotoxins, Toxicon 38 (2) (2000) 245–258, https://doi.org/10.1016/S0041-0101(99)00153-1.

[31] M. Cui, S. Khanijou, J. Rubino, K.R. Aoki, Subcutaneous administration of botulinum toxin a reduces formalin-induced pain, Pain 107 (1–2) (2004) 125–133, https://doi.org/10.1016/j.pain.2003.10.008.

[32] M. Schueler, W.L. Neuhuber, R. De Col, K. Messlinger, Innervation of rat and human dura mater and pericranial tissues in the parieto-temporal region by meningeal afferents, Headache J. Head Face Pain 54 (6) (2014) 996–1009, https://doi.org/10.1111/head.12371.

[33] B. Kosaras, M. Jakubowski, V. Kainz, R. Burstein, Sensory innervation of the calvarial bones of the mouse, J. Comp. Neurol. 515 (3) (2009) 331–348, https://doi.org/10.1002/cne.22049.

[34] M. Schueler, K. Messlinger, M. Dux, W.L. Neuhuber, R. De, Extracranial projections of meningeal afferents and their impact on meningeal nociception and headache, Pain 154 (9) (2013) 1622–1631, https://doi.org/10.1016/j.pain.2013.04.040.

[35] R. Noseda, A. Melo-Carrillo, R.R. Nir, A.M. Strassman, R. Burstein, Non-trigeminal nociceptive innervation of the posterior dura: Implications to occipital headache, J. Neurosci. 39 (10) (2019) 1867–1880, https://doi.org/10.1523/JNEUROSCI.2153-18.2018.

[36] R. Burstein, A.M. Blumenfeld, S.D. Silberstein, A.M. Adams, M.F. Brin, Mechanism of action of OnabotulinumtoxinA in chronic migraine: a narrative review, Headache J. Head Face Pain 60 (7) (2020) 1259–1272, https://doi.org/10.1111/head.13849.

[37] A. Blumenfeld, S.D. Silberstein, D.W. Dodick, S.K. Aurora, C.C. Turkel, W.J. Binder, Method of injection of OnabotulinumtoxinA for chronic migraine: a safe, well-tolerated, and effective treatment paradigm based on the PREEMPT clinical program, Headache J. Head Face Pain 50 (9) (2010) 1406–1418, https://doi.org/10.1111/j.1526-4610.2010.01766.x.

[38] N.T. Mathew, B.M. Frishberg, M. Gawel, R. Dimitrova, J. Gibson, C. Turkel, Botulinum toxin type A (botox ®) for the prophylactic treatment of chronic daily headache: a randomized, double-blind, placebo-controlled trial, Headache J. Head Face Pain 45 (4) (2005) 293–307, https://doi.org/10.1111/j.1526-4610.2005.05066.x.

[39] S.D. Silberstein, S.R. Stark, S.M. Lucas, S.N. Christie, R.E. DeGryse, C.C. Turkel, Botulinum toxin type A for the prophylactic treatment of chronic daily headache: a randomized, double-blind, placebo-controlled trial, Mayo Clin. Proc. 80 (9) (2005) 1126–1137, https://doi.org/10.4065/80.9.1126.

[40] F.G. Freitag, S. Diamond, M. Diamond, U. George, Botulinum toxin type A in the treatment of chronic migraine without medication overuse, Headache J. Head Face Pain 48 (2) (2008) 201–209, https://doi.org/10.1111/j.1526-4610.2007.00963.x.

[41] D.W. Dodick, C.C. Turkel, R.E. DeGryse, S.K. Aurora, S.D. Silberstein, R.B. Lipton, H.-C. Diener, M.F. Brin, OnabotulinumtoxinA for treatment of chronic migraine: pooled results from the double-blind, randomized, placebo-controlled phases of the PREEMPT clinical program, Headache J. Head Face Pain 50 (6) (2010) 921–936, https://doi.org/10.1111/j.1526-4610.2010.01678.x.

[42] M. Lanteri-Minet, A. Ducros, C. Francois, E. Olewinska, M. Nikodem, L. Dupont-Benjamin, Effectiveness of onabotulinumtoxinA (BOTOX®) for the preventive treatment of chronic migraine: a meta-analysis on 10 years of real-world data, Cephalalgia 42 (14) (2022) 1543–1564, https://doi.org/10.1177/03331024221123058.

[43] S. Ashina, G.M. Terwindt, T.J. Steiner, M.J. Lee, F. Porreca, C. Tassorelli, T.J. Schwedt, R.H. Jensen, H.- Christoph Diener, R.B. Lipton, Medication overuse headache, Nat. Rev. Dis. Prim. 9 (1) (2023), https://doi.org/10.1038/s41572-022-00415-0.

[44] A. Hovaguimian, J. Roth, Management of chronic migraine, BMJ (2022) e067670, https://doi.org/10.1136/bmj-2021-067670.

[45] C.S. Dhanasekara, D. Payberah, J.Y. Chyu, C.L. Shen, C.N. Kahathuduwa, The effectiveness of botulinum toxin for chronic tension-type headache prophylaxis: a systematic

review and meta-analysis, Cephalalgia 43 (3) (2023), https://doi.org/10.1177/03331024221150231.
[46] R. Anton, G. Juodzbalys, The use of botulinum toxin A in the management of trigeminal neuralgia: a systematic literature review, J. Oral Maxillofac. Res. 11 (2) (2020), https://doi.org/10.5037/jomr.2020.11202.
[47] W.J. Becker, Botulinum toxin in the treatment of headache, Toxins 12 (12) (2020) 803, https://doi.org/10.3390/toxins12120803.
[48] S. Santos-Lasaosa, M.L. Cuadrado, A.B. Gago-Veiga, A.L. Guerrero-Peral, P. Irimia, J.M. Láinez, R. Leira, J. Pascual, J. Porta-Etessam, M. Sánchez del Río, J. Viguera Romero, P. Pozo-Rosich, Evidencia y experiencia del uso de onabotulinumtoxinA en neuralgia del trigémino y cefaleas primarias distintas de la migraña crónica, Neurologia 35 (8) (2020) 568−578, https://doi.org/10.1016/j.nrl.2017.09.003.
[49] A.A. Argyriou, D.D. Mitsikostas, E. Mantovani, M. Vikelis, S. Tamburin, Beyond chronic migraine: a systematic review and expert opinion on the off-label use of botulinum neurotoxin type-A in other primary headache disorders, Taylor and Francis Ltd., Greece Exp. Rev. Neurotherapeut. 21 (8) (2021) 923−944, https://doi.org/10.1080/14737175.2021.1958677.
[50] L. Bendtsen, S. Sacco, M. Ashina, D. Mitsikostas, F. Ahmed, P. Pozo-Rosich, P. Martelletti, Guideline on the use of onabotulinumtoxinA in chronic migraine: a consensus statement from the European Headache Federation, J. Headache Pain 19 (1) (2018) 91, https://doi.org/10.1186/s10194-018-0921-8.
[51] M.T. Corasaniti, G. Bagetta, P. Nicotera, A. Tarsitano, P. Tonin, G. Sandrini, G.W. Lawrence, D. Scuteri, Safety of onabotulinumtoxin A in chronic migraine: a systematic review and meta-analysis of randomized clinical trials, Toxins 15 (5) (2023) 332, https://doi.org/10.3390/toxins15050332.
[52] H.T. Wong, M. Khalil, F. Ahmed, OnabotulinumtoxinA for chronic migraine during pregnancy: a real world experience on 45 patients, J.Headache Pain 21 (1) (2020), https://doi.org/10.1186/s10194-020-01196-1.
[53] M.F. Brin, R.S. Kirby, A. Slavotinek, M.A. Miller-Messana, L. Parker, I. Yushmanova, H. Yang, Pregnancy outcomes following exposure to onabotulinumtoxinA, Pharmacoepidemiol. Drug Safety 25 (2) (2016) 179−187, https://doi.org/10.1002/pds.3920.
[54] Drugs and Lactation Database. OnabotulinumtoxinA, (2006) 2006.
[55] C.J. Vaughn, Drugs and lactation database: LactMed, J. Electron. Resour. Med. Libr. 9 (4) (2012) 272−277, https://doi.org/10.1080/15424065.2012.735134.
[56] R. Marcelo, B. Freund, The efficacy of botulinum toxin in pediatric chronic migraine: a literature review, J. Child Neurol. 35 (12) (2020) 844−851, https://doi.org/10.1177/0883073820931256.
[57] P.K. Winner, M. Kabbouche, M. Yonker, V. Wangsadipura, A. Lum, M.F. Brin, A randomized trial to evaluate OnabotulinumtoxinA for prevention of headaches in adolescents with chronic migraine, Headache J. Head Face Pain 60 (3) (2020) 564−575, https://doi.org/10.1111/head.13754.
[58] S. Shah, M.D. Calderon, N. Crain, J. Pham, J. Rinehart, Effectiveness of onabotulinumtoxinA (BOTOX) in pediatric patients experiencing migraines: a randomized, double-blinded, placebo-controlled crossover study in the pediatric pain population, Regio. Anesthes. Pain Med. 46 (1) (2021) 41−48, https://doi.org/10.1136/rapm-2020-101605.
[59] A.M. Blumenfeld, R.J. Stark, M.C. Freeman, A. Orejudos, A. Manack Adams, Long-term study of the efficacy and safety of OnabotulinumtoxinA for the prevention of chronic migraine: COMPEL study, J. Headache Pain 19 (1) (2018), https://doi.org/10.1186/s10194-018-0840-8.
[60] C. Altamura, R. Ornello, F. Ahmed, A. Negro, A.M. Miscio, A. Santoro, A. Alpuente, A. Russo, M. Silvestro, S. Cevoli, N. Brunelli, L. Grazzi, C. Baraldi, S. Guerzoni, A.P. Andreou, G. Lambru, I. Frattale, K. Kamm, R. Ruscheweyh, M. Russo, P. Torelli, E. Filatova, N. Latysheva, A. Gryglas-Dworak, M. Straburzynski, C. Butera, B. Colombo, M. Filippi, P. Pozo-Rosich, P. Martelletti, S. Sacco, F. Vernieri, OnabotulinumtoxinA in elderly patients with chronic migraine: insights

from a real-life European multicenter study, J. Neurol. 270 (2) (2023) 986–994, https://doi.org/10.1007/s00415-022-11457-5.

[61] S.K. Aurora, P. Winner, M.C. Freeman, E.L. Spierings, J.O. Heiring, R.E. DeGryse, A.M. VanDenburgh, M.E. Nolan, C.C. Turkel, OnabotulinumtoxinA for treatment of chronic migraine: pooled analyses of the 56-week PREEMPT clinical program, Headache J. Head Face Pain 51 (9) (2011) 1358–1373, https://doi.org/10.1111/j.1526-4610.2011.01990.x.

[62] E. Cernuda-Morollón, C. Ramón, D. Larrosa, R. Alvarez, N. Riesco, J. Pascual, Long-term experience with onabotulinumtoxinA in the treatment of chronic migraine: what happens after one year? Cephalalgia 35 (10) (2015) 864–868, https://doi.org/10.1177/0333102414561873.

[63] I. Aicua-Rapun, E. Martínez-Velasco, A. Rojo, A. Hernando, M. Ruiz, A. Carreres, E. Porqueres, S. Herrero, F. Iglesias, A.L. Guerrero, Real-life data in 115 chronic migraine patients treated with Onabotulinumtoxin A during more than one year, J. Headache Pain 17 (1) (2016), https://doi.org/10.1186/s10194-016-0702-1.

[64] A.M. Blumenfeld, G. Kaur, A. Mahajan, H. Shukla, K. Sommer, A. Tung, K.L. Knievel, Effectiveness and safety of chronic migraine preventive treatments: a systematic literature review, Pain Therap. 12 (1) (2023) 251–274, https://doi.org/10.1007/s40122-022-00452-3.

[65] P.K. Winner, A.M. Blumenfeld, E.J. Eross, A.C. Orejudos, D.L. Mirjah, A.M. Adams, M.F. Brin, Long-term safety and tolerability of OnabotulinumtoxinA treatment in patients with chronic migraine: results of the COMPEL study, Drug Safety 42 (8) (2019) 1013–1024, https://doi.org/10.1007/s40264-019-00824-3.

[66] A. Masters-Israilov, M.S. Robbins, OnabotulinumtoxinA wear-off phenomenon in the treatment of chronic migraine, Headache 59 (10) (2019) 1753–1761, https://doi.org/10.1111/head.13638.

[67] A. Zidan, C. Roe, D. Burke, L. Mejico, OnabotulinumtoxinA wear-off in chronic migraine, observational cohort study, J. Clin. Neurosci. 69 (2019) 237–240, https://doi.org/10.1016/j.jocn.2019.07.043.

[68] A. Negro, M. Curto, L. Lionetto, P. Martelletti, A two years open-label prospective study of OnabotulinumtoxinA 195 U in medication overuse headache: a real-world experience, J. Headache Pain 17 (2016) 2016.

[69] S. Choudhury, M.R. Baker, S. Chatterjee, H. Kumar, Botulinum toxin: an update on pharmacology and newer products in development, Toxins 13 (1) (2021), https://doi.org/10.3390/TOXINS13010058.

Chapter | Three

Occipital nerve blocks for headaches

Ahish Chitneni[1], Joe Ghorayeb[2], Sandeep Yerra[3], Alan David Kaye[4]

[1]Department of Rehabilitation and Regenerative Medicine, New York-Presbyterian Hospital - Columbia and Cornell, New York, NY, United States; [2]University of Medicine and Health Sciences, New York, NY, United States; [3]Montefiore Health Systems, Bronx, NY, United States; [4]Department of Anesthesiology, Louisiana State University Health Sciences Center, Shreveport, LA United States

INTRODUCTION

Targeting the occipital nerves through an anesthetic block is utilized both as a diagnostic and therapeutic procedure for commonly encountered headaches. With over 48% of the general population reporting experiencing headaches and approximately 1.4%–2.2% of the global population experiencing headaches lasting at least 2 weeks per month this painful disorder has the capacity to reduce one's quality of life, resulting in significant disability and socioeconomic burden.

The importance of occipital nerve blocks (ONBs) for the management of headache syndromes is derived from understanding the origin, anatomy, and course of the occipital nerves and how dysfunction, irritation, and/or compression of these nerves may contribute to headache development. The occipital nerves originate from the cervical spinal nerves C2and C3. They consist of the greater occipital nerve (GON), the lesser occipital nerve (LON), and the third occipital nerve (TON). All three nerves are located in the posterior neck and scalp regions and are interconnected through their communicating branches. The primary function of these nerves is to provide the sensory supply to the skin overlying the posterior and lateral scalp, including the skin of the external ear. The occipital nerves mainly carry sensory fibers with only the TON carrying some motor innervation to the semispinalis capitis muscle.

Greater occipital nerve (GON)

The GON is the largest afferent nerve that arises from the medial division of the dorsal ramus of the C2 spinal nerve. It runs between the C1 and C2 vertebrae

Interventional Management of Migraines and Other Headache Disorders. https://doi.org/10.1016/B978-0-443-23557-3.00003-6

and courses between the inferior capitis oblique and semispinalis capitis muscles from underneath the suboccipital triangle. The GON pierces several muscles, including the semispinalis capitis, trapezius, and inferior oblique, before reaching the subcutaneous layer. Due to its complex involvement with surrounding muscles, the GON can potentially be compressed, entrapped, or irritated. After piercing the fibrous layer of the trapezius and sternocleidomastoid, the GON reaches the scalp and superior nuchal line, where it provides innervation to the skin at the back of the scalp, vertex of the skull, ear, and above the parotid gland.

LESSER OCCIPITAL NERVE (LON)

The LON originates from the ventral rami of the C2 and C3 spinal nerves and courses to the occipital region along the posterior margin of the sternocleidomastoid muscle. It pierces the deep cervical fascia close to the cranium and travels to the cephalad thereafter. Near the cranium, it penetrates the deep cervical fascia and traverses superiorly above the occiput to innervate the skin and communicate with the GON. The LON has three branches: the auricular, mastoid, and occipital branches. The LON innervates the scalp in the lateral region of the head behind the ear and the cranial surface of the ear.

THIRD OCCIPITAL NERVE (TON)

The TON is a superficial medial branch of the dorsal ramus of the C3 spinal nerve. The TON travels through the dorsolateral surface of the C2–C3 facet joint. Tubbs et al. found that the TON sends out small branches that travel across the midline and interact with the contralateral TON in 66.7% of individuals. The TON perforates the splenius capitis, trapezius, and semispinalis capitis. The TON communicates with the GON and innervates the region of the skin below the superior nuchal line after innervating the semispinalis capitis.

INDICATIONS

Clinically, occipital nerve blocks (ONB) are commonly utilized for the treatment of cervicogenic headaches, cluster headaches, and occipital neuralgia with supportive evidence reported in the scientific literature for the treatment of migraine. While study characteristics do raise questions regarding the permanence of outcomes along with uncertainty in regard to patient selection criteria to predict outcome success with this intervention, the available data support the utility of ONBs for a variety of headache syndromes.

TYPES OF HEADACHES TREATED WITH ONB INJECTIONS

Cervicogenic headache

A cervicogenic headache presents as unilateral pain that starts in the neck and is referred from bony structures or soft tissues of the neck. It is a common chronic and recurrent headache that usually starts following neck movement and is often accompanied by reduced range of motion (ROM) of the neck.

The diagnostic criteria for cervicogenic headache must include the following:

1 The source of the pain must be in the neck and perceived in the head or face.
2 Evidence that the pain may be attributed to arising from the neck must be confirmed by way of clinical signs that implicate a source of pain in the neck or abolition of a headache following diagnostic blockade of a cervical structure or its nerve supply using a placebo or other adequate controls.
3 The pain resolves within 3 months after successful treatment of the causative disorder or lesion

The anatomic basis and pathophysiologic mechanisms explaining the convergence of the upper cervical and trigeminal sensory pathways in addition to the bidirectional referral of nociception between the neck and trigeminal receptive fields of the head and face leading to the referral of cervicogenic headache from a cervical source to the forehead, temple, or orbit form the foundation for managing cervicogenic headache through ONB.

Randomized controlled trials have demonstrated that administering ONBs results in a significant decrease in pain scores and rescue analgesics consumption.

Inan et al. compared the effect of GON blocks to C2/C3 spinal rami blocks in 28 patients with cervicogenic headaches and concluded that both blocks are equally effective. No significant difference was observed between the two groups in terms of pain frequency or degree of pain, except for pain frequency in the first week following the first therapeutic block, which was significantly reduced in the C2/C3 group.

Naja et al. evaluated 50 patients with cervicogenic headaches who received GON and LON blocks with or without facial nerve blocks. The anesthetic block group, which received a mixture of lidocaine, bupivacaine, epinephrine, fentanyl, and clonidine was compared with the placebo group, which received normal saline, and realized a statistically significant improvement in pain intensity, frequency, and duration as well as a decrease in analgesic use at 2 weeks when compared with the placebo group.

Lauretti et al. evaluated 30 patients with cervicogenic headaches and compared the efficacy of classical and subcompartmental injection techniques. The authors concluded that the suboccipital compartmental GON technique resulted in at least 24 weeks compared with 2 weeks of analgesia when the same dosage of dexamethasone and lidocaine was applied by the classical technique. This suggests that the administration of the drugs closer to the dorsal ganglion was more efficacious to counteract cervicogenic headache symptoms.

Pingree et al. conducted a noncontrolled prospective trial where they evaluated 14 patients who underwent ultrasound-guided GON blocks at the C2 level and reported a successful block in 86% of patients at 30 minutes postinjection. A significant decrease in the mean numerical rating scale (NRS) score was observed at 30 minutes, 2 weeks, and 4 weeks compared with baseline.

Ertem and Yilmaz retrospectively evaluated 21 patients with cervicogenic headache who underwent at least three GON blocks and attended at least three follow-up appointments. A significant reduction in pain scores was reported at 3 months posttreatment.

CLUSTER HEADACHE

Cluster headaches are the most common of the primary headache type known as trigeminal autonomic cephalgias. Although rare, they have earned the reputation as one of the most severe types of headaches. A genetic basis regarding the acquisition of cluster headaches is appreciated, though the mode of inheritance remains unclear. Cluster headaches are characterized as short-lasting, unilateral headaches with at least one accompanying autonomic sign or symptom ipsilateral to the headache such as lacrimation, nasal congestion (not to be confused with sinus headache), conjunctival injection, or aural fullness. Cluster headaches may occur at varying frequencies from every other day to eight times a day. They typically occur at approximately the same time of day, most often at night. Most patients experience episodic headaches, with daily attacks lasting for weeks to months, followed by remission for months to years.

In a randomized-controlled trial (RCT) conducted by Ambrosini et al. which evaluated the effect of a single suboccipital injection of local anesthetics and corticosteroids in individuals with cluster headaches, the study participants received lidocaine 2% (0.5 mL) with either betamethasone or saline (control group). Cluster headache attacks were diminished in 61% of individuals in the lidocaine + betamethasone group within 72 hours for up to 4 weeks when compared with those in the control group who did not report sustained relief.

A retrospective study Peres et al. evaluated the effect of a single greater occipital nerve block (GONB) on 14 patients with cluster headaches. Nerve blockade was achieved using 3 mL of lidocaine 1% with triamcinolone 40 mg. Nine patients (64%) reportedly achieved a good or moderate response following the intervention over a period of 3–7 days.

OCCIPITAL NEURALGIA

Reflected in the third edition of the International Classification of Headache Disorders (ICHD-3), occipital neuralgia is described as a unilateral or bilateral paroxysmal, shooting, or stabbing pain in the posterior part of the scalp, in the distribution(s) of the greater, lesser, and/or third occipital nerves, sometimes accompanied by diminished sensation or dysesthesia in the affected area and commonly associated with tenderness over the involved nerve(s).

A case series by Vanderhoek et al. involving two patients with occipital neuralgia compared the difference in outcomes between the administration of bilateral GONBs alone and bilateral GONBs in addition to bilateral GON pulsed radiofrequency ablation (PRFA). Both patients responded immediately after the intervention and the benefit persisted for several months in the patient who received both GONB and PRFA.

In a retrospective study by Kastler et al. analyzing the effect of CT-guided GON-block in 33 patients with occipital neuralgia, pain reduction >50% was observed in 86% of patients with a mean pain relief duration of 9 months (range 3–24).

MIGRAINE HEADACHE

Migraine is a common disabling primary headache disorder. In the Global Burden of Disease Study 2010 (GBD2010), it was ranked as the third most prevalent disorder in the world. In GBD2015, it was ranked third–highest cause of disability worldwide in both males and females under the age of 50 years.

Several RCTs have been conducted to determine the efficacy of ONB for migraine treatment. Ashkenazi et al. sought to learn whether adding triamcinolone to local anesthetics increased the efficacy of GONB and trigger-point injections (TPI) in individuals with transformed migraine. Thirty-seven patients were included in the study (18 in the control group and 19 in the treatment group). The authors concluded that while both groups experienced significant rapid relief within 20 minutes following the intervention, the addition of triamcinolone to local anesthetics did not result in improved clinical outcomes at 4 weeks follow-up.

Similar findings were noted by Kashipazha et al., who determined that GONB with two regimens including either triamcinolone in combination with lidocaine ($n = 24$) or normal saline with lidocaine ($n = 24$) resulted in considerable pain reduction, severity and frequency as well as analgesic use for up to 2 months after the intervention. Though no significant differences were attributed to the difference in drug regimens.

Inan et al. aimed to assess the efficacy of GONB in patients with chronic migraine treatment. Eighty-four patients were randomized into two groups. GONB was administered four times (once per week) with saline in the control group or bupivacaine in the treatment group. After 4 weeks of treatment, blinding was removed and the control group was administered 4 weekly injections with bupivacaine, while the treatment group continued to receive bupivacaine once per month. The primary endpoint was the difference in number of headache days, duration of headache, and pain scores. The authors concluded that bupivacaine was superior to placebo at 1 month ($P < .001$) and this effect was sustained at 3 months follow-up ($P < .001$).

Friedman et al. sought to determine whether GONB was as effective as intravenous metoclopramide for acute migraine treatment in a double-dummy, double-blind, parallel-arm, noninferiority study conducted in two emergency departments. Ninety-nine patients with migraine of moderate or severe intensity were randomized to receive either bilateral GONB with each side administered bupivacaine 0.5% 3 mL ($n = 51$), or metoclopramide 10 mg IV ($n = 48$). Patients who received the GONB reported a mean improvement of 5.0 (95% CI: 4.1, 5.8) while those who received metoclopramide reported a larger mean improvement of 6.1 (95% CI: 5.2, 6.9). Sustained headache relief was reported by 11/51 (22%) GONB and 18/47 (38%) metoclopramide patients. Of the 51 GONB patients, 17 (33%) required rescue medication in the ED versus 8/48 (17%) metoclopramide patients. An adverse event was reported by 16/51 (31%) GONB patients and 18/48 (38%) metoclopramide patients. The authors concluded that while GONB was not as efficacious as intravenous metoclopramide, an exploratory analysis suggested that in experienced hands, GONB and intravenous metoclopramide may have comparable efficacy.

PATIENT SELECTION CRITERIA

In various studies, occipital nerve tenderness to palpation (TTP) or reproduction of headache pain with ON pressure (RHPONP) was employed as a selection criterion based on the rationale that their presence implicates the ON or its downstream connections as part of the pain generator.

Afridi et al. found that tenderness around the region of the GON was significantly associated with a positive response to the injection in patients with primary headache syndromes.

For cervicogenic headaches, the utility of ON TTP or RHPONP to appropriately select candidates for ONB is inconclusive. Naja et al. and Anthony found positive results with RHPONP, however, Inan et al., Vincent et al., and Bovim and Sand did not use either method as a selection criterion and also obtained positive results.

For cluster headaches, Anthony et al. used RHPONP as a selection criterion and obtained generally positive results, but several others did not use either RHPONP or ON TTP as a selection criterion and also obtained positive results.

By definition, occipital neuralgia involves TTP of the ONS and Anthony et al. obtained positive results using this selection criterion.

Whether RHPONP or ON TTP predicts success in migraine treatment remains unclear. Two studies using RHPONP as a selection criterion were Anthony et al. and Tobin et al., both of which obtained positive results. Anthony et al. and Saadah and Taylor used ON TTP as selection criteria and obtained positive results. Bovim and Sand used neither ON TTP nor RHPONP and achieved positive outcomes.

Therefore, the current body of evidence suggests that these selection criteria are not necessary for cervicogenic, cluster headache or migraine syndromes. More research is necessary to determine which selection criteria, if any, are suitable for predicting positive outcomes with ONBs.

TECHNIQUE

ONBs consist of targeting the greater occipital nerve, lesser occipital nerve, and the third occipital nerve to provide analgesia for a range of conditions as discussed in previous sections. Each nerve provides sensory innervation to various parts of the occipital region as seen in Fig. 3.1 and can be targeted based on the clinical presentation of the patient undergoing treatment.

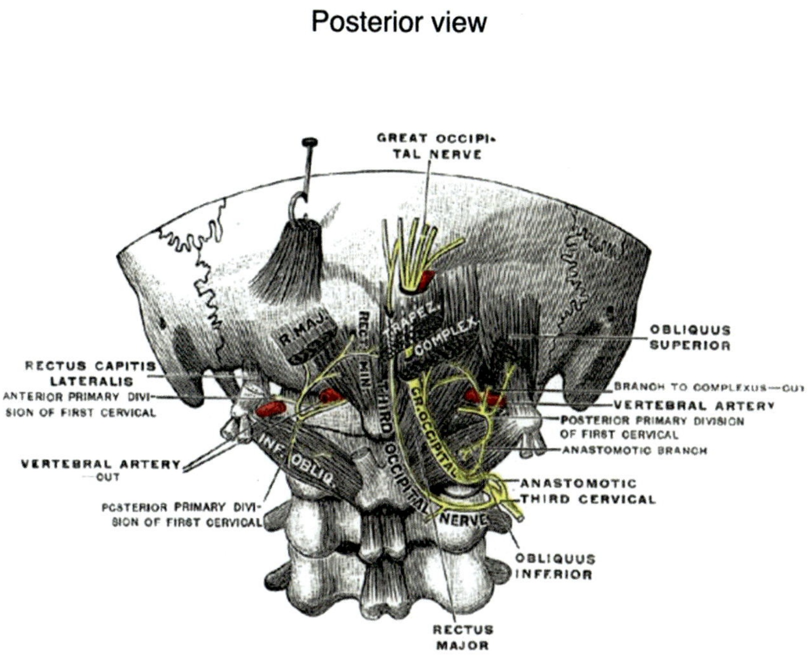

FIGURE 3.1 Nerves of the occipital region. Courtesy of Craig Hacking, Radiopaedia.org, rID: 96129.

All three procedures, greater occipital nerve block, lesser occipital nerve block, and third occipital nerve block can be conducted both through ultrasound guidance and fluoroscopic guidance.

GREATER OCCIPITAL NERVE BLOCK

Ultrasound technique

In the ultrasound-guided technique of the greater occipital nerve block, the patient positioning includes a sitting position with cervical flexion as seen in Fig. 3.2.

First step involves the palpation of the nuchal ridge and placement of a 6–13 MHz linear transducer probe over the area between the C1 lateral mass and the C2 spinous process. Additionally, several techniques note the importance of palpation of the occipital artery at the site of the superior nuchal ridge to confirm location. After the placement of the linear transducer probe, the use of Doppler can be used to confirm location of the occipital artery. The location of the GON is typically near the occipital artery and can be seen as a hypoechoic ovoid shaped structure. Next, a 25-gauge, 1.5 inch needle is advanced in an in-plane approach toward the occipital nerve until reaching the periosteum. During this part of the procedure, patients may note paresthesias at the location of sensory distribution of the GON as seen in Fig. 3.1 After proper identification of the nerve, as seen in Fig. 3.3, aspiration is conducted to ensure

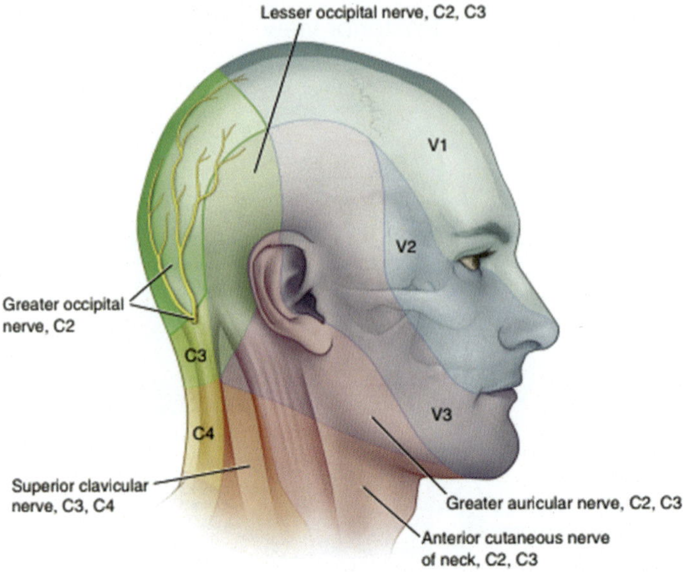

FIGURE 3.2 Sensory distribution of the greater and lesser occipital nerve.

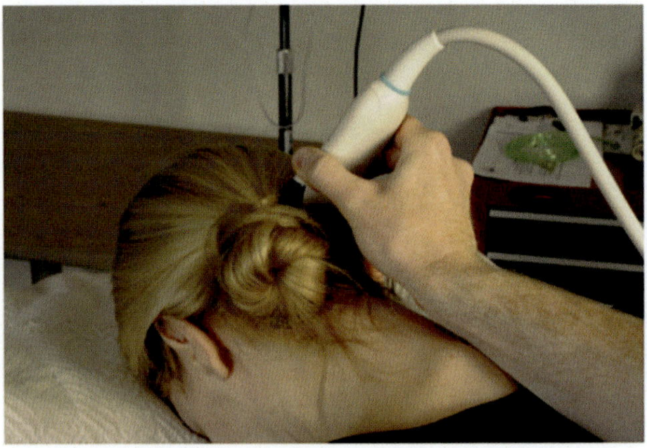

FIGURE 3.3 Patient positioning during greater occipital nerve block.

extravascular location of the needle and an injectate containing 5 mL of anesthetic (0.5% bupivacaine or ropivacaine) with a steroid is injected in a fanlike manner around the nerve.

After injecting, the needle is removed and pressure is placed at the site of the injection to avoid bleeding and hematoma formation. Various other identification techniques for the GON have noted that the identification of the semispinalis capitis muscle (SsCM) and the obliquus capitis inferior muscle may be important landmarks to identify the GON as it typically is found ascending on the dorsal surface of the OCIM as seen in Fig. 3.4.

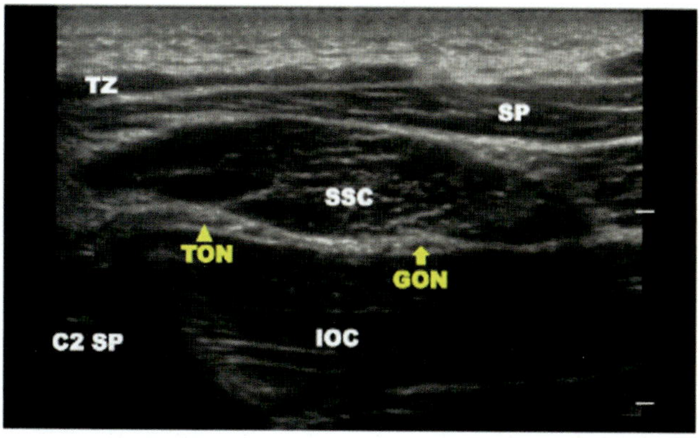

FIGURE 3.4 Identification of the greater occipital nerve on ultrasound.

Fluoroscopy technique

During the fluoroscopic technique for this procedure, the patient positioning consists of a sitting position with cervical flexion as seen in Fig. 3.2 In this procedure, 3 cm lateral to the occipital protuberance, the skin entry point is marked. At the site of the entry point, a 25-gauge 1.5 inch needle is inserted perpendicularly in a posteroanterior manner. As seen in Fig. 3.1, the needle is advanced to the location of the GON. After aspiration is conducted to ensure extravascular location of the needle and an injectate containing 5 mL of anesthetic (0.5% bupivacaine or ropivacaine) with a steroid is injected in a fanlike manner.

LESSER OCCIPITAL NERVE BLOCK

Ultrasound technique

In the ultrasound-guided technique for a lesser occipital nerve (LON) block, the patient positioning includes the patient in a lateral recumbent position as a 6–13 MHz linear transducer probe is positioned in a transverse manner over the posterior border of the SCM. As the transducer probe is placed in this location, the LON can be visualized as a hypoechoic ovoid structure in the posterior margin of the SCM as seen in Fig. 3.5.

Next, a 25-gauge, 1.5-inch needle is advanced from posterior to anterior in an in-plane approach above the LON at the superficial layer of the SCM in the posterior margin. After proper identification of the location, aspiration is conducted to ensure extravascular location of the needle and an injectate containing 5 mL of anesthetic (0.5% bupivacaine or ropivacaine) with dexamethasone or betamethasone is injected in a fanlike manner around the nerve.

Fluoroscopy technique

For the fluoroscopic technique for a LON Block, the patient positioning includes a sitting position with cervical flexion. For this procedure, the greater

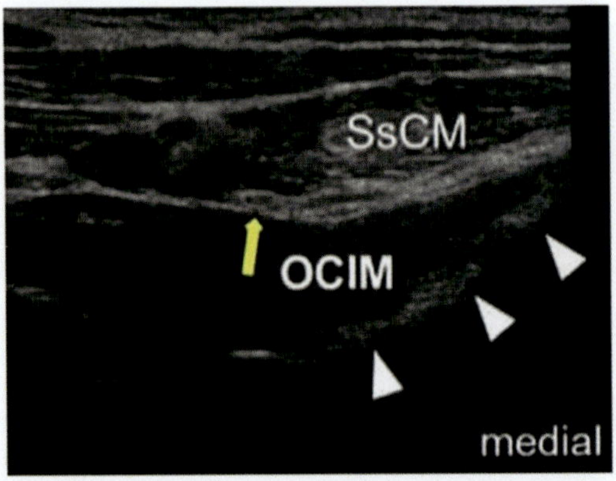

FIGURE 3.5 Identification of the GON (*Yellow* line) between the semispinalis capitis muscle (SsCM) and obliquus capitis inferior muscle (OCIM).

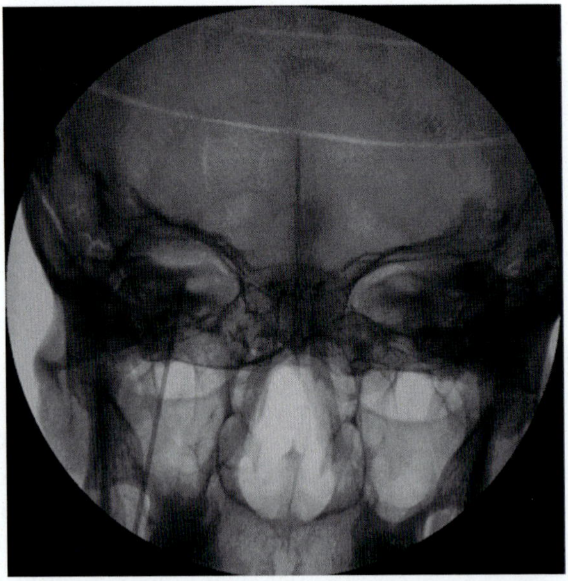

FIGURE 3.6 Fluoroscopic view of the needle at the location of the greater occipital nerve.

occipital artery is palpated at the superior nuchal ridge and the area latero-inferior to the artery is marked as the entry site. A 25-gauge, 1.5 inch needle is used to advance toward the periosteum while using fluoroscopic imaging as guidance. Fig. 3.6 demonstrates the needle positioning for targeting the LON on fluoroscopy. After aspiration is conducted to ensure extravascular location of the needle and an injectate containing 5 mL of anesthetic (0.5% bupivacaine or ropivacaine) with a steroid is injected in a fanlike manner.

COMBINED GREATER OCCIPITAL NERVE AND LESSER OCCIPITAL NERVE BLOCK WITH ONE INJECTION SITE

Another method of injection suggests the possibility of a combined GON and LON block with one injection site. In this method, ultrasound guidance is used with the patient positioned seated with cervical flexion. Next, the superior nuchal ridge is identified while palpating the greater occipital artery. Ultrasound guidance can be used with the Doppler to identify the greater occipital artery and confirm location. Next, a 25-gauge, 1.5 inch needle is introduced between the GON and LON and advanced to the level of the periosteum. After reaching the periosteum, the needle is advanced medially toward the GON and 2 mL of the injectate (local anesthetic + steroid) is injected. After injection, the needle is withdrawn and redirected to the mastoid process to the location of the LON.

THIRD OCCIPITAL NERVE BLOCK

Ultrasound technique

For conducting an ultrasound-guided block of the Third Occipital Nerve (TON), the patient is positioned in a lateral recumbent position. Next, the mastoid process is palpated and the ultrasound probe is placed at the inferior border of the mastoid process. Next, the ultrasound probe is moved anterior to posterior to the mastoid until visualization of the transverse process of C1. After visualization of the C1 transverse process, the C2 transverse process is identified by slight rotation of the transducer. After identification of the C2 transverse process, the transducer is moved posteriorly to identify both the C2–C3 articulation as well as the C3–C45 articulation. In between the articulation of these two joints, the medial branch of the C3 joint can be visualized as well as the TON can be visualized crossing the C2–C3 articulation 1 mm from the bone as seen in Fig. 3.1 After identification of the nerve, a 25-gauge, 1.5-inch needle is introduced in an out-of-plane approach to approach the TON. After aspiration is conducted to ensure extravascular location of the needle and an injectate containing 5 mL of anesthetic (0.5% bupivacaine or ropivacaine) with a steroid is injected in a fanlike manner.

Fluoroscopy technique

For the fluoroscopic technique of a TON block, patient positioning includes a prone position with cervical spine flexion. Next, imaging of the C2–C3 facet joint with a lateral fluoroscopic image is obtained. Target for the TON will be a rectangular area between the anterior edge of the superior articular process of C3 as seen in Fig. 3.7.

Next, a 25-gauge 1.5-inch needle is used with 2% lidocaine for local anesthetic at the site of entry. After local anesthesia is administered, a 22-gauge 3.5-inch spinal needle is advanced through the entry point to engage the periosteum. After engaging the periosteum, the needle is withdrawn 3 mm to ensure that there is no contact with the facet joint. After proper placement, contrast is

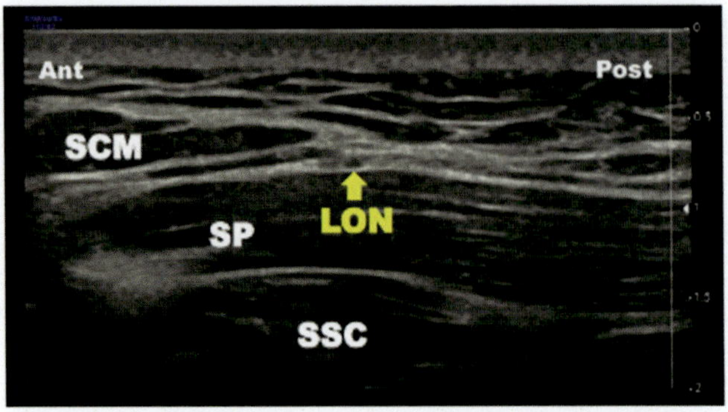

FIGURE 3.7 Localizing of the lesser occipital nerve (LON) at the posterior margin of the SCM.

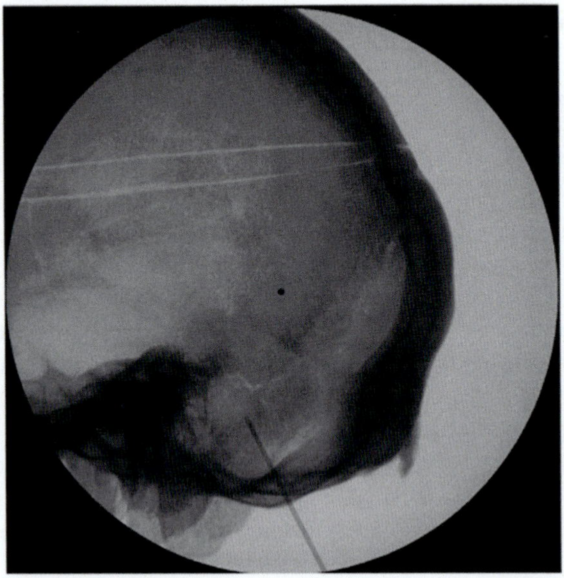

FIGURE 3.8 Needle positioning for targeting the lesser occipital nerve (LON).

injected to observe a periscapular flow around the C2–C3 joint to ensure that the needle is in the correct location as seen in Fig. 3.8.

After confirmation with contrast, the injectate consisting of 1 mL local anesthetic and a steroid is injected and observed under fluoroscopic guidance.

EFFICACY AND SAFETY

ONB's (occipital nerve blocks) are often employed as the first choice in the treatment of occipital neuralgia (ON). As with spinal blocks they can be both therapeutic and diagnostic.

Occipital headaches are often difficult to diagnose, ONB's serve a crucial role in differentiating ON from cervicogenic headaches The treatment paradigm of cervicogenic headaches is quite different from ON's. Cervicogenic headaches require neuroimaging and employ different treatment options such as zygapophyseal joint blocks, radiofrequency techniques etc. The diagnostic role of ONB's is widely accepted with the recent inclusion of ONB's as of the diagnostic criteria by the international Classification of Headache Disorders, third edition (ICHD-3)

In regards to the therapeutic role of ONB's, the efficacy is quite high with minimal systemic side effects. Juskys et al., in his study showed a response rate of 95% at 6 months with a significant decrease of mean VAS (visual analogue scale of five points). Moreover, the effects are immediate providing relief within 24 hours. The same study also showed the need for adjuvant analgesic medication decreased from 100% before the pain procedure to 17% after the block. The blocking of GON versus LON or third occipital nerve is highly provider and patient dependent, sometimes requiring bilateral blocks. The procedure can be repeated many times as the systemic effects are minimal with local blocks of ONB's. ONB's in appropriate cases are used in 4-week intervals to wean patients off the pain medications. In patients with successful blocks there might be a role for RFA's and PNS but research is limited to small studies, which are showing positive results. In addition, people who receive incomplete or partial relief can be supplemented with oral agents.

Most of the side effects of ONB's seems to be local rather than systemic. They also are mostly procedure-related, which seems to be self-resolving in nature. Local neuralgia, neck pain, injection site pain, vertigo and dizziness are the most common side effects. Per Chowdary et al., in a meta-analysis of 226 studies found minimal systemic side effects with zero cases of treatment adverse side effects in all the RCT's. Systemic side effects of the steroids are reported in a few case reports and are likely related to the addition of steroids The problem remains that in acute attacks of ONs, which only last from seconds to minutes, ONB's cannot be used as a rescue procedure. The issue is that this requires an experienced pain provider to perform the procedure. The other obvious complication is procedural where one can injure the occipital artery, which can be prevented either through palpation or the use of USG/Doppler. Limited evidence exists for the addition of steroids with LA's in ONB's However, addition of steroids seems to provide longer lasting effects.

The main challenge remains isolation of the GON nerve, which is primarily done through anatomical identification in clinical practice. As with any anatomical identification it is highly variable and provider dependent. However, there have been recent studies that seem to advocate for USG guided isolation of GON with positive results.

In conclusion ONB's remain a cornerstone in diagnosis and treatment of ON's with ONB's serving as the first line of treatment with minimal side effects. Further studies involving local anesthesia, steroids, and additives are currently ongoing to best determine the most effective inject for these treatments (Figs. 3.9–3.11).

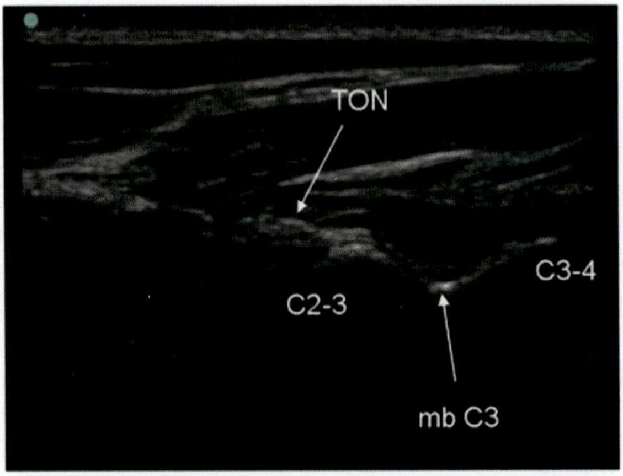

FIGURE 3.9 Please check the sentence "The safety and efficacy…" for clarity.

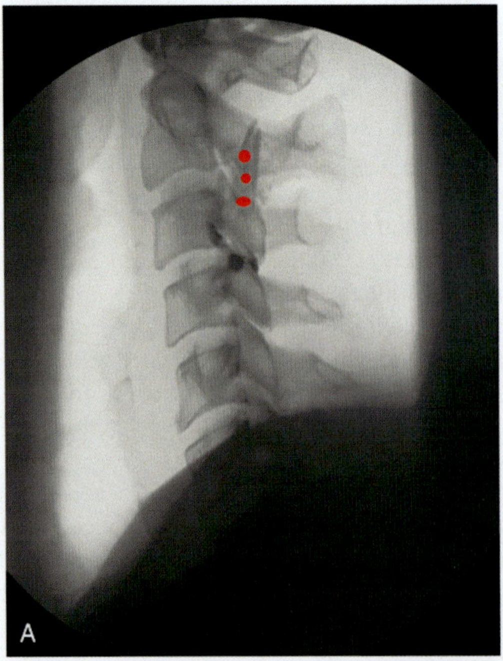

FIGURE 3.10 Rectangular Area Marking the location of the TON Target.

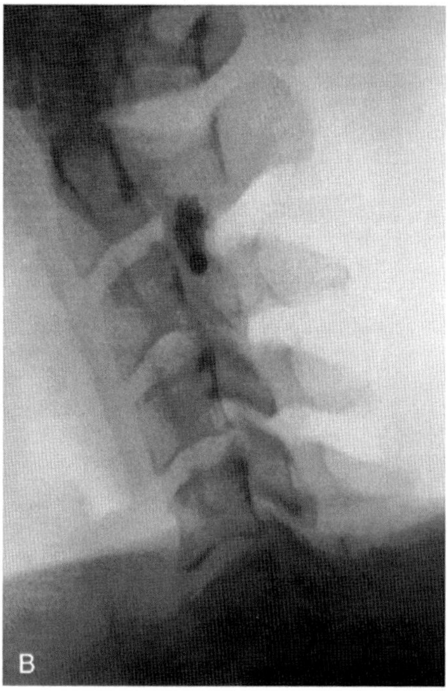

FIGURE 3.11 Contrast over the TON.

FURTHER READING

[1] Headache classification committee of the international headache society (IHS) the international classification of headache disorders, 3rd edition, Cephalalgia 38 (1) (2018) 1–211, https://doi.org/10.1177/0333102417738202.

[2] A. Blumenfeld, A. Ashkenazi, U. Napchan, S.D. Bender, B.C. Klein, R. Berliner, et al., Expert consensus recommendations for the performance of peripheral nerve blocks for headaches—a narrative review, Headache 53 (3) (2013) 437–446.

[3] T.J. Mungoven, L.A. Henderson, N. Meylakh, Chronic migraine pathophysiology and treatment: a review of current perspectives, Front. Pain Res. (Lausanne) 2 (2021) 705276, https://doi.org/10.3389/fpain.2021.705276.

[4] B.R. Hasırcı Bayır, G. Gürsoy, C. Sayman, G.A. Yüksel, Y. Çetinkaya, Greater occipital nerve block is an effective treatment method for primary headaches?. Primer bas ağrılarında büyük oksipital sinir blokajı etkin bir tedavi yöntemi mi? Agri 34 (1) (2022) 47–53, https://doi.org/10.14744/agri.2021.32848.

[5] J.H. VanderPluym, L. Charleston 4th, M.E. Stitzer, C.C. Flippen 2nd, C.E. Armand, J. Kiarashi, A review of underserved and vulnerable populations in headache medicine in the United States: challenges and opportunities, Curr. Pain Headache Rep. 26 (6) (2022) 415–422, https://doi.org/10.1007/s11916-022-01042-w.

[6] W.J. Kemp 3rd, R.S. Tubbs, A.A. Cohen-Gadol, The innervation of the scalp: a comprehensive review including anatomy, pathology, and neurosurgical correlates, Surg. Neurol. Int. 2 (2011) 178, https://doi.org/10.4103/2152-7806.90699.

[7] I. Choi, S.R. Jeon, Neuralgias of the head: occipital neuralgia, J. Kor. Med. Sci. 31 (4) (2016) 479–488, https://doi.org/10.3346/jkms.2016.31.4.479.

[8] R.S. Tubbs, M.M. Mortazavi, M. Loukas, et al., Anatomical study of the third occipital nerve and its potential role in occipital headache/neck pain following midline dissections

of the craniocervical junction, J. Neurosurg. Spine 15 (1) (2011) 71–75, https://doi.org/10.3171/2011.3.SPINE10854.
[9] O. Sjaastad, C. Saunte, H. Hovdahl, H. Breivik, E. Grønbaek, "Cervicogenic" headache. An hypothesis, Cephalalgia 3 (4) (1983) 249–256, https://doi.org/10.1046/j.1468-2982.1983.0304249.x.
[10] A.W. Kane, D.S. Diaz, C. Moore, Physical therapy management of adults with mild traumatic brain injury, Semin. Speech Lang. 40 (1) (2019) 36–47, https://doi.org/10.1055/s-0038-1676652.
[11] B. Wu, L. Yue, F. Sun, S. Gao, B. Liang, T. Tao, The feasibility and efficacy of ultrasound-guided C2 nerve root coblation for cervicogenic headache, Pain Med. 20 (6) (2019) 1219–1226, https://doi.org/10.1093/pm/pny227.
[12] L.S. Moye, A.F. Tipton, I. Dripps, et al., Delta opioid receptor agonists are effective for multiple types of headache disorders, Neuropharmacology 148 (2019) 77–86, https://doi.org/10.1016/j.neuropharm.2018.12.017.
[13] C. Mares, J.H. Dagher, M. Harissi-Dagher, Narrative review of the pathophysiology of headaches and photosensitivity in mild traumatic brain injury and concussion, Can. J. Neurol. Sci. 46 (1) (2019) 14–22, https://doi.org/10.1017/cjn.2018.361.
[14] N. Bogduk, Cervicogenic headache: anatomic basis and pathophysiologic mechanisms, Curr. Pain Headache Rep. 5 (4) (2001) 382–386, https://doi.org/10.1007/s11916-001-0029-7.
[15] N. Inan, A. Ceyhan, L. Inan, O. Kavaklioglu, A. Alptekin, N. Unal, C2/C3 nerve blocks and greater occipital nerve block in cervicogenic headache treatment, Funct. Neurol. 16 (3) (2001) 239–243.
[16] Z.M. Naja, M. El-Rajab, M.A. Al-Tannir, F.M. Ziade, O.M. Tawfik, Occipital nerve blockade for cervicogenic headache: a double-blind randomized controlled clinical trial, Pain Pract. 6 (2) (2006) 89–95, https://doi.org/10.1111/j.1533-2500.2006.00068.x.
[17] G.R. Lauretti, S.W. Corrêa, A.L. Mattos, Efficacy of the greater occipital nerve block for cervicogenic headache: comparing classical and subcompartmental techniques, Pain Pract. 15 (7) (2015) 654–661, https://doi.org/10.1111/papr.12228.
[18] M.J. Pingree, J.S. Sole, T.G. O' Brien, J.S. Eldrige, S.M. Moeschler, Clinical efficacy of an ultrasound-guided greater occipital nerve block at the level of C2, Reg. Anesth. Pain Med. 42 (1) (2017) 99–104, https://doi.org/10.1097/AAP.0000000000000513.
[19] D.H. Ertem, I. Yilmaz, The effects of repetitive greater occipital nerve blocks on cervicogenic headache, Turk. J. Neurol. 25 (2019).
[20] P.D. Drummond, Mechanisms of autonomic disturbance in the face during and between attacks of cluster headache, Cephalalgia 26 (6) (2006) 633–641, https://doi.org/10.1111/j.1468-2982.2006.01106.x.
[21] J. Hoffmann, A. May, Diagnosis, pathophysiology, and management of cluster headache, Lancet Neurol. 17 (1) (2018) 75–83, https://doi.org/10.1016/S1474-4422(17)30405-2.
[22] J. Weaver-Agostoni, Cluster headache, Am. Fam. Physic. 88 (2) (2013) 122–128.
[23] M.J. Láinez, E. Guillamón, Cluster headache and other TACs: pathophysiology and neurostimulation options, Headache 57 (2) (2017) 327–335, https://doi.org/10.1111/head.12874.
[24] A. Ambrosini, M. Vandenheede, P. Rossi, et al., Suboccipital injection with a mixture of rapid- and long-acting steroids in cluster headache: a double-blind placebo-controlled study, Pain 118 (1–2) (2005) 92–96, https://doi.org/10.1016/j.pain.2005.07.015.
[25] M.F. Peres, M.A. Stiles, H.C. Siow, T.D. Rozen, W.B. Young, S.D. Silberstein, Greater occipital nerve blockade for cluster headache, Cephalalgia 22 (7) (2002) 520–522, https://doi.org/10.1046/j.1468-2982.2002.00410.x.
[26] https://ichd-3.org/wp-content/uploads/2018/01/The-International-Classification-of-Headache-Disorders-3rd-Edition-2018.pdf.
[27] M.D. Vanderhoek, H.T. Hoang, B. Goff, Ultrasound-guided greater occipital nerve blocks and pulsed radiofrequency ablation for diagnosis and treatment of occipital neuralgia, Anesth Pain Med. 3 (2) (2013) 256–259, https://doi.org/10.5812/aapm.10985.

[28] A. Kastler, Y. Onana, A. Comte, A. Attyé, J.L. Lajoie, B. Kastler, A simplified CT-guided approach for greater occipital nerve infiltration in the management of occipital neuralgia, Eur. Radiol. 25 (8) (2015) 2512–2518, https://doi.org/10.1007/s00330-015-3622-6.
[29] T. Vos, A.D. Flaxman, M. Naghavi, et al., Years lived with disability (YLDs) for 1160 sequelae of 289 diseases and injuries 1990-2010: a systematic analysis for the Global Burden of Disease Study 2010, Lancet 380 (9859) (2012) 2163–2196, https://doi.org/10.1016/S0140-6736(12)61729-2.
[30] T.J. Steiner, L.J. Stovner, T. Vos, Gbd 2015: migraine is the third cause of disability in under 50s, J. Headache Pain 17 (1) (2016) 104, https://doi.org/10.1186/s10194-016-0699-5.
[31] A. Ashkenazi, R. Matro, J.W. Shaw, M.A. Abbas, S.D. Silberstein, Greater occipital nerve block using local anaesthetics alone or with triamcinolone for transformed migraine: a randomised comparative study, J. Neurol. Neurosurg. Psychiatr. 79 (4) (2008) 415–417, https://doi.org/10.1136/jnnp.2007.124420.
[32] D. Kashipazha, A. Nakhostin-Mortazavi, S.E. Mohammadianinejad, M. Bahadoram, S. Zandifar, S. Tarahomi, Preventive effect of greater occipital nerve block on severity and frequency of migraine headache, Glob. J. Health Sci. 6 (6) (2014) 209–213, https://doi.org/10.5539/gjhs.v6n6p209.
[33] L.E. Inan, N. Inan, Ö. Karadas, et al., Greater occipital nerve blockade for the treatment of chronic migraine: a randomized, multicenter, double-blind, and placebo-controlled study, Acta Neurol. Scand. 132 (4) (2015) 270–277, https://doi.org/10.1111/ane.12393.
[34] B.W. Friedman, E. Irizarry, A. Williams, et al., A randomized, double-dummy, emergency department-based study of greater occipital nerve block with bupivacaine vs intravenous metoclopramide for treatment of migraine, Headache 60 (10) (2020) 2380–2388, https://doi.org/10.1111/head.13961.
[35] S.K. Afridi, K.G. Shields, R. Bhola, P.J. Goadsby, Greater occipital nerve injection in primary headache syndromes–prolonged effects from a single injection, Pain 122 (1–2) (2006) 126–129, https://doi.org/10.1016/j.pain.2006.01.016.
[36] M. Anthony, Cervicogenic headache: prevalence and response to local steroid therapy, Clin. Exp. Rheumatol. 18 (2 Suppl. 19) (2000) S59–S64.
[37] M.B. Vincent, R.A. Luna, D. Scandiuzzi, S.A. Novis, Greater occipital nerve blockade in cervicogenic headache, Arq. Neuropsiquiatr. 56 (4) (1998) 720–725, https://doi.org/10.1590/s0004-282x1998000500004.
[38] G. Bovim, T. Sand, Cervicogenic headache, migraine without aura and tension-type headache. Diagnostic blockade of greater occipital and supra-orbital nerves, Pain 51 (1) (1992) 43–48, https://doi.org/10.1016/0304-3959(92)90007-X.
[39] J.A. Tobin, S.S. Flitman, Occipital nerve blocks: effect of symptomatic medication: overuse and headache type on failure rate, Headache 49 (10) (2009) 1479–1485, https://doi.org/10.1111/j.1526-4610.2009.01549.x.
[40] M. Anthony, Headache and the greater occipital nerve, Clin. Neurol. Neurosurg. 94 (4) (1992) 297–301, https://doi.org/10.1016/0303-8467(92)90177-5.
[41] H.A. Saadah, F.B. Taylor, Sustained headache syndrome associated with tender occipital nerve zones, Headache 27 (4) (1987) 201–205, https://doi.org/10.1111/j.1526-4610.1987.hed2704201.x.
[42] S.D. Waldman, Greater occipital nerve block, in: Atlas of Pain Management Injection Techniques, fifth ed., Elsevier, Amsterdam, 2022, pp. 76–80.
[43] H.T. Benzon, et al., Head and neck blocks, in: Practical Management of Pain, sixth ed., Elsevier, Philadelphia, PA, 2023, pp. 857–873.
[44] Nysora, Ultrasound-Guided Greater Occipital Nerve Block, NYSORA, 30 November 2022. www.nysora.com/pain-management/ultrasound-guided-greater-occipital-nerve-block.
[45] Z. Song, S. Zhao, J. Ma, Z. Wu, S. Yang, Fluoroscopy-guided blockade of the greater occipital nerve in Cadavers: a comparison of spread and nerve involvement for different injectate volumes, Pain Res. Manag. 2020 (September 22, 2020) 8925895, https://doi.org/10.1155/2020/8925895.

[46] G.C. Fowler, Procedures to treat headaches, in: Pfenninger & Fowler's Procedures for Primary Care, Elsevier, Philadelphia, PA, 2020, pp. 59–67.
[47] S.D. Waldman, Lesser occipital nerve block, in: Atlas of Pain Management Injection Techniques, fifth ed., Elsevier, Amsterdam, 2022, pp. 81–85.
[48] S.D. Waldman, Third occipital nerve block, in: Atlas of Pain Management Injection Techniques, fifth ed., Elsevier, Amsterdam, 2022, pp. 86–90.
[49] A. Siegenthaler, et al., Ultrasound-guided third occipital nerve and cervical medial branch nerve blocks, Tech. Reg. Anesth. Pain Manag. 13 128–132.
[50] C.J. Williams, et al., Cervical injection techniques, in: Atlas of Interventional Orthopedics Procedures: Essential Guide for Fluoroscopy and Ultrasound-Guided Procedures, Elsevier, London, 2023, pp. 134–165.
[51] A. Ashkenazi, M. Levin, Three common neuralgias. How to manage trigeminal, occipital, and postherpetic pain, Postgrad. Med. 116 (3) (2004), https://doi.org/10.3810/pgm.2004.09.1579.
[52] J.B. Ward, Greater occipital nerve block, Semin. Neurol. 23 (1) (2003) 59–62, https://doi.org/10.1055/s-2003-40752.
[53] R. Juškys, G. Šustickas, Effectiveness of treatment of occipital neuralgia using the nerve block technique: a prospective analysis of 44 patients, Acta Med. Litu. 25 (2) (2018) 53–60, https://doi.org/10.6001/actamedica.v25i2.3757.
[54] D. Chowdhury, D. Datta, A. Mundra, Role of greater occipital nerve block in headache disorders: a narrative review, Neurol. India 69 (Suppl. ment) (2021) S228–S256, https://doi.org/10.4103/0028-3886.315993.
[55] T.A. Seeger, S. Orr, L. Bodell, L. Lockyer, T. Rajapakse, K.M. Barlow, Occipital nerve blocks for pediatric posttraumatic headache: a case series, J. Child Neurol. 30 (9) (2015) 1142–1146, https://doi.org/10.1177/0883073814553973.
[56] K.G. Shields, M.J. Levy, P.J. Goadsby, Alopecia and cutaneous atrophy after greater occipital nerve infiltration with corticosteroid, Neurology 63 (11) (2004) 2193–2194, https://doi.org/10.1212/01.wnl.0000145832.26051.3c.
[57] R.C. Tripathi, S.K. Parapuram, B.J. Tripathi, Y. Zhong, K.V. Chalam, Corticosteroids and glaucoma risk, Drugs Aging 15 (6) (1999) 439–450, https://doi.org/10.2165/00002512-199915060-00004.
[58] F. Dach, Á.L. Éckeli, S. Ferreira Kdos, J.G. Speciali, Nerve block for the treatment of headaches and cranial neuralgias - a practical approach, Headache 55 (Suppl. 1) (2015) 59–71, https://doi.org/10.1111/head.12516.

Chapter | Four

Trigger point injections

Sohyun Kang[1], Tsan-Chen Yeh[2], Anish Rana[3], Jeremy Tuttle[3,4], Alan David Kaye[5,6]

[1]Department of Rehabilitation Medicine, Columbia University Medical Center, Weill Cornell Medical College, NewYork-Presbyterian Hospital, New York, NY, United States; [2]California University of Science and Medicine, Colton, CA, United States; [3]University of Rochester Medical Center, Rochester, NY, United States; [4]Weill Cornell Medical College, New York, NY, United States; [5]Department of Anesthesiology, Louisiana State University Health Sciences Center, Shreveport, LA, United States; [6]LSU Health Shreveport (Formerly LSU Health Sciences Center Shreveport), Department of Anesthesiology, Shreveport, LA, United States

INTRODUCTION

Trigger point injection (TPI) is a common type of therapeutic modality used to treat pain from a variety of musculoskeletal and neurologic disorders. Trigger points (TPs) are localized areas of skeletal muscle with taut bands that are commonly observed in myofascial pain. These points are usually palpable as focal knots and produce a characteristic referred pain upon palpation. Trigger points may be formed due to repeated stress on the muscle fibers from trauma or repeated micro-trauma [1].

The pathophysiology behind TPs is poorly understood but the hypothesis is that they form due to abnormal endplate potentials that in turn lead to excessive acetylcholine release in the neuromuscular junction. This process leads to the formation of a taut band from the sustained muscle contraction, which leads to local ischemia, release of inflammatory substances, and peripheral sensitization [2]. In a study by Shat et al., patients with active TPs had elevated concentrations of various pain mediators such as substance P, calcitonin gene-related peptide, bradykinin, serotonin and other neurochemicals compared with those who had latent TPs or healthy controls [3].

TPIs have demonstrated effectiveness as a therapeutic intervention for treating pain, especially as part of a multidisciplinary treatment program that includes physical therapy. TPIs can inactivate trigger points by causing a temporary relaxation of the taut muscle cord and allowing improved perfusion. By targeting specific points of myofascial pain, TPIs can directly treat the pathologic tissue by addressing the patient's pain generator breaking the pain cycle, which can improve patients' range of motion and overall, functionally. TPIs are

Interventional Management of Migraines and Other Headache Disorders. https://doi.org/10.1016/B978-0-443-23557-3.00004-8

relatively safe with little to no side effects and the growing accessibility of ultrasound has improved available options for image guided TPIs [1].

INDICATION FOR TPI

TPs in the head and neck areas have been associated with various primary and secondary headache disorders and TPIs can target the mechanical sites that evoke or perpetuate an underlying headache disorder. The evidence for TPIs include randomized controlled trials for tension-type headache and migraine, with the strongest evidence for frequent tension-type headache [4]. Studies have found that TPs are found more often in patients with chronic tension-type headache, and the presence of active TPs was also associated with greater intensity and longer duration of headache [5]. The most common sites of injection are the trapezius, sternocleidomastoid, and temporalis muscles, but other muscles of the head and neck are also commonly injected [6]. In a 2010 survey of American Headache Society (AHS) membership, 75.3% of responders reported performing TPIs in the treatment of headache. They are commonly performed in concert with peripheral nerve blocks [7].

Tension-type headache

Tension-type headache (TTH) is the most prevalent type of primary headache, affecting about 40% of people with headaches [8]. Active TPs at head and neck muscles, including the suboccipital, upper trapezius, sternocleidomastoid, temporalis, superior oblique, and lateral rectus muscles, have been associated with both episodic and chronic tension-type headaches [6]. Multiple studies have been conducted to investigate the effectiveness of TP inactivation in TTH management. In a double-blinded, placebo-controlled randomized study done by Karadas et all, repetitive lidocaine injection at pericranial muscles has shown improvements in headache symptoms and duration in patients with frequent episodic TTH when compared with placebo (saline) or single injection group [9]. In another double-blind, placebo-controlled randomized study, Karadas et al. investigated the therapeutic effect of TPI in chronic TTH. Though administered with other nonpharmacological interventions such as peripheral nerve block, TPI injection with lidocaine had again been associated with greater improvements in headache severity and duration among other symptoms when compared with saline injection [10].

Migraine

While migraine is recognized as a neurovascular disorder that is predominantly associated with activation of the trigeminal-vascular system, localized scalp tenderness and referred pain have been observed in migraine patients [11]. In head and neck muscles, a greater number of TPS has been identified in subjects with migraine than in healthy controls [5,12]. While the association between the number of active TPs and migraine duration or intensity remains unclear, multiple studies have been done to study the efficacy of TPIs in migraine management.

In the study done by Ashkenazi and Young, the effect of greater occipital nerve block (GONB) with and without TPIs was examined in treating head pain and allodynia in migraine subjects. Acute reduction in headache pain was observed in 89.5% of patients (most received TPIs), and the symptoms of allodynia were also reduced according to the questionnaire in the study [13]. In another study, the effectiveness of TPIs with ropivacaine was evaluated as a prophylactic treatment in patients with severe migraine. One or more TPs have been identified in all subjects, mostly near the temporal and/or suboccipital areas. Ropivacaine 10 mg was injected weekly for 12 weeks in 52 subjects, of whom 65% had chronic severe migraine and 53% had medication overuse. Frequency of migraine attacks was reduced in 31 (59.6%) patients after 12 weeks. Decreased severity of the attacks was also reported by the subjects, in addition to reduction of rescue medication intake in 11 (21%) subjects [14]. As migraine remains to be one of the most common indications for TPIs, the association between TPs and migraine duration intensity and intensity should not be undermined. The efficacy of TPIs both as an acute and prophylactic migraine management should be further examined.

Cluster headache

Active TPs have also been associated with cluster headaches. In a study done by Calandre et al., TPs were found in all 12 patients [15]. In the study, anesthetics were injected as abortive, preemptive, and prophylactic therapy, all of which showed high success rates (85%, 86%, 88%). In 7 out of eight patients with chronic cluster headaches, TPIs have also been shown to decrease disease severity and frequency when combined with prophylactic drug therapy. While the study was uncontrolled and limited by its sample size, the role of TPs inactivation remains to be a great potential treatment option for refractory cluster headache management.

Cervicogenic headache

In an investigation done by Jaeger, more trigger points have been identified on the symptomatic side of cervicogenic headache (CGH) patients compared with the asymptomatic side. Noninvasive pain management treatments focused on the myofascial points have been shown to decrease headache frequency and intensity, suggesting that TPs might be an important source of pain in CGH patients [16].

Multiple studies have also suggested an association between active TPs at sternocleidomastoid muscle and CGH [17]. A case report has identified SCM MtP to be the primary source of headache pain in a CGH patient [18]. In another preliminary study done by Gema Bodes-Pardo et al., TP therapy on sternocleidomastoid active muscle trigger points in CGH has been shown to decrease headache intensity and neck pain while increasing cervical range of motion [19].

Lastly, in a retrospective study done by Baron et al., the role of trigemino-cervical and vestibular circuitry was investigated in producing the cervically-mediated symptoms in CGH, including dizziness, tinnitus, headache, and ear discomfort. After receiving greater occipital nerve block (GONB) and TPI, improvements were seen in 57% of patients with headaches, along with symptomatic improvement in dizziness (46%), neck range of motion (71%), ear discomfort (47%), and tinnitus (30%) [20]. The result suggests the potential benefit of TPIs in managing CGH symptoms, and further study would be needed.

Posttraumatic headache

While no specific study has been done to study the effectiveness of TPI in treating posttraumatic headaches (PTH) [19,21], myofascial trigger points have been associated with posttraumatic headaches. In a study done by Castaldo et al., trigger points were found in neck muscle in both whiplash-associated injury and mechanical neck pain, suggesting that TPIs of local anesthetics given at neck muscles might be helpful, along with other interventional treatment, might be an appropriate future treatment option for PTH [22].

Overall, while there is a lack of hard evidence on the efficacy of TPIs in treating headache disorders, the associations between TPs and both primary and secondary headache disorders suggest an important role TPIs might play in managing headache disorders. Further studies on specific medication selection, injection sites, and the overall effectiveness of TPIs are needed.

PATIENT SELECTION CRITERIA

The general indication for trigger point injections technique is myofascial pain syndrome with symptomatic active trigger points. The key to treating headache disorders with injection technique is thus to identify myofascial trigger points in the muscle, ligament, and tendon of the head and neck area. The examiner should carefully examine the trapezius, sternocleidomastoid, splenius, semispinalis capitis, semispinalis cervicis, temporalis muscles, and the trochlear, masseter, and temporomandibular areas. When pressed, active trigger points at the head and neck area can elicit referred pain patterns similar to the patient's main headache complaints. A careful physical and neurological exam should also be conducted to exclude other structural or neurological causes of headaches [21].

It is crucial to recognize contraindications before performing TPIs to maximize patient safety, prevent any adverse complications, and optimize patient outcomes. Absolute contraindications for a TPI include the presence of active infection over trigger point site, local open skull defect beneath the intended injection site, and inaccessibility of trigger point by needle. Relative contraindications include, but are not limited to, patients on anticoagulation medications, pregnant patients, patients with obscure anatomical landmarks, patients with a history of severe fibromyalgia or keloid formation, or patients with a history of local anesthetic allergy. For patients with a history of bleeding disorder and anticoagulation use, it is recommended to monitor labs closely before the procedure and practice safe techniques such as compressing the injection sites 5–10 min after the injection. In the pregnant population, the type of anesthetics used should also be carefully monitored and usually lidocaine (FDA Category B) is preferred over bupivacaine (FDA Category C). For patients with ambiguous anatomical landmarks, TPIs could be performed with electromyographic or ultrasound guidance [21,23].

TECHNIQUE FOR TRIGGER POINT INJECTIONS

TPI has shown effectiveness in delivering rapid and symptomatic relief through deactivation of TPs [24]. Generally, this procedure involves the precise injection of medication, typically local anesthetics, directly into specific localized trigger points [21]. Despite widespread use of trigger point injections as treatment for headaches and migraines, there seems to be a lack of standardization in methodology among practitioners utilizing TPIs [7,25]. However, there is a recommended routine procedure that can be followed to ensure consistency and optimal outcomes.

The proper equipment for a TPI procedure include, but are not limited to the following: 27- to 30-gauge 1.5-inch needle, 3, 5, or 10-mL syringe, gloves, local anesthetic agent such as lidocaine or bupivacaine, alcohol pads for disinfecting the skin, blood pressure monitor and pulse oximeter, and other optional injectable solutions such as sodium bicarbonate, topical anesthetic spray, corticosteroids, and steroids [23,24].

In order to prepare a patient for TPI, it is imperative to obtain informed consent and ensure the availability of necessary equipment. Depending on the specific location of the trigger point, the patient may be positioned either in a seated or recumbent posture, with the latter being preferable to mitigate the risk of vasovagal reactions. Adequate disinfection of the patient is essential, and this can be achieved by cleansing the area with an alcohol solution [21,23]. The selection of needles should be based on the desired depth of penetration, with superficial injections requiring a 22-, 25-, or 27-gauge, 1.5-inch needle, while deeper injections necessitate the use of a 21-, 2.5-inch needle [21,24].

After ensuring appropriate patient preparation and equipment setup, the clinician should correctly identify and firmly stabilize the trigger point using a pinch grip between the index finger and thumb of the nondominant hand [23]. Prior to needle injection, the clinician should notify the patient that they may experience an unpleasant sensation or sharp pain upon needle contact with the trigger point [24]. The needle will be inserted approximately one–1.5 cm away from the trigger point, gradually advancing at a 30-degree angle into the targeted trigger point. The clinician should perform needle aspiration to verify that there is no contact between the needle and any vascular structure, thereby preventing the systemic distribution of the anesthetic. 0.1−0.3 mL of 1% lidocaine or 0.5% bupivacaine should then be injected into the trigger point. Administering additional injections in the vicinity of the primary TP might be beneficial, with a recommended total volume ranging from 2 to 4 mL [21]. The volumes of trigger point injections may vary from one patient to another. Clinicians have the option to use ultrasound-guided injection for greater accuracy and to avoid neurovascular structures [23]. Following a TPI, postinjection soreness might result; re-injection is not recommended until this soreness resolves (typically after 3−4 days) [24].

EFFICACY

The findings from the reviewed studies indicate that TPIs can provide significant relief for certain types of headaches, such as tension-type headaches and migraine headaches. In several randomized controlled trials, TPIs were found to be superior to placebo injections in reducing headache frequency, intensity, and duration.

In a 108-participant double-blind randomized controlled trial Karadas et al. studied the use of TPIs in episodic tension-type headaches (ETTH). Participants were randomized into four groups: (1) a single injection of normal saline, (2) a single injection of lidocaine 0.5%, (3) five injections of normal saline administered on alternate days, and (4) five injections of lidocaine 0.5% administered on alternate days. TPIs were shown to have significant reduction on the visual analogue scale (VAS) and frequency of painful days per month (FPD). Both lidocaine groups outperformed the comparable normal saline groups at 2- and 4-month time intervals in terms of the frequency of FPD and VAS and visual analogue scale ratings ($P < .05$). Notably, only the group receiving five lidocaine injections exhibited sustained improvement at the 6 month follow-up. Interestingly, this study also demonstrated that the groups receiving five injections, regardless of the injectate used, achieved better outcomes compared with the groups receiving only one injection. This may point to the value in repeated TPIs in those suffering from ETTHs. Of note, there was no difference between the frequency of painful days months after the treatment ($P > .005$). This suggests TPIs efficacy in the short-term relief of headaches, while it is not a cure [9].

In a separate 48-participant double-blind RCT, Karadas et al. compared efficacy of lidocaine to normal saline in TPIs for relieving pain in patients with chronic tension-type headaches (CTTH). Participants received TPIs with either normal saline or lidocaine 0.5% every 3 days for a total of three injections. Evaluation of the treatment outcomes took place 3 months after the completion of injections. Both groups exhibited significant reductions in various parameters, including the number of painful days, values on the visual analogue scale, monthly analgesic consumption, scores on the Hamilton Depression Rating Scale, and scores on the Hamilton Anxiety Rating Scale. However, the lidocaine group demonstrated superior response to treatment across all measures compared with the normal saline group, with a statistically significant difference of $P < .001$. This discrepancy could potentially be attributed to the increased injection frequency and the specific targeting of TPIs, which encompassed not only particular muscle groups but also injections at the exit points of the fifth cranial nerve and around the superior cervical ganglion. It is worth noting that inadvertent nerve blockade at these sites may have influenced the finding that lidocaine yielded superior outcomes in comparison to normal saline [10]. Despite the promising findings, it is important to acknowledge the limitations of the current evidence. Many studies included small sample sizes and lacked long-term follow-up, making it challenging to draw definitive conclusions about the sustained efficacy of TPIs. Furthermore, the heterogeneity in study designs, patient populations, and outcome measures complicates the comparison of results across studies.

COMMON ADVERSE EFFECTS AND MANAGEMENT

While trigger point injections are generally considered safe, some potential adverse effects can occur. The most common adverse effects include injection site pain or discomfort, bruising, bleeding, and infection. These adverse effects are usually mild and self-limiting, but they can also be prevented by carefully avoiding vascular structures and areas that appear inflamed. Patients may also experience vasovagal reaction to the procedure, in which case the physician should have the patient in a secure position and remain to minimize the risk of a fall injury [26].

Although rare, serious adverse effects may occur with TPIs. These include nerve injury, allergic reactions to the injected medications, pneumothorax (injections in the upper back), and vascular injury [27]. Some considerations to prevent direct nerve damage include using a small needle (25 gauge or smaller), performing straight needle movements, and avoiding lateral movements. Symptoms of nerve injury to look out for include new onset neuropathic pain, numbness, or paresthesia. To prevent a pneumothorax, for example, the physician should move the tissue away from the intercostal space to negate the risk of lung injury [28]. Proper technique and anatomical knowledge are crucial to minimize the risk of these complications [23].

To manage common adverse effects, patients can be advised to apply ice packs to the injection site and take over-the-counter pain relievers, if necessary. Reassurance and counseling about the expected duration and nature of the adverse effects are important in promoting patient comfort and compliance.

In case of rare but serious adverse effects, immediate medical attention should be sought. Nerve injuries may require referral to a specialist, such as a neurologist or neurosurgeon, for further evaluation and management. Allergic reactions should be managed according to established protocols, including administration of antihistamines and, in severe cases, epinephrine.

Pneumothorax and vascular injuries require prompt recognition and appropriate intervention by a trained healthcare professional [28].

PATIENT OUTCOMES AND SATISFACTION

In terms of headache frequency, many studies have demonstrated a reduction in the number of headaches experienced per month. Patients have reported a decrease in the frequency of both episodic and chronic headaches, with some experiencing a significant decrease in the number of headache days per month. Additionally, TPIs have been shown to be effective in reducing the intensity and duration of headaches. Patients have reported experiencing less severe and shorter-lasting headaches after the procedure. Moreover, TPIs have been found to alleviate associated symptoms commonly experienced with headaches, such as neck and shoulder pain, sensitivity to light and sound, and nausea. By targeting trigger points and relieving muscle tension, TPIs can provide relief not only for the primary headache but also for the secondary symptoms associated with it [21].

Patient satisfaction is an important aspect to consider when evaluating the effectiveness of any medical procedure. Studies assessing patient satisfaction with trigger point injections for headache treatment have consistently reported high levels of satisfaction among patients. Many patients express relief and gratitude for the improvement in their headache symptoms and the impact it has on their quality of life [29].

Patients often appreciate the relatively quick and straightforward nature of the procedure. TPIs can usually be performed in an outpatient setting, with minimal downtime or recovery period. The immediate relief experienced by some patients further contributes to their satisfaction with the procedure. The ability of TPIs to provide rapid relief, especially during acute headache episodes, is highly valued by patients. Furthermore, the noninvasive nature of TPIs compared with more invasive interventions, such as nerve blocks or surgery, is often seen as a favorable aspect by patients. The procedure typically involves a simple injection into the trigger points without the need for incisions or extensive recovery. By injecting anesthetics directly into the identified trigger point, TPIs provide localized relief while minimizing potential systemic side effects [21].

There are also important disadvantages of TPIs that need to be considered. In addition to the potential adverse side effects discussed above, the relief provided by TPIs are often temporary and do not address the underlying cause of the trigger point. If the TPs are secondary to underlying medical conditions, repeated injections and lifestyle changes are often necessary for long-term relief. A thorough history with physical and neurological exams are thus necessary to identify the cause of patients' symptoms.

Patient satisfaction may also be influenced by other factors, such as the healthcare provider's communication, the overall treatment plan, and the patient's expectations. Proper patient education and counseling about the potential benefits and limitations of TPIs can help manage expectations and enhance patient satisfaction.

TPIs have shown promising results in the management of headaches, with improvements reported in headache frequency, intensity, duration, and associated symptoms. Patients consistently express high levels of satisfaction with the procedure, attributing their satisfaction to the relief experienced, the convenience of the procedure, and its noninvasive nature. However, further research with larger sample sizes and longer follow-up periods is warranted to strengthen the evidence base and fully understand the long-term outcomes and patient satisfaction associated with trigger point injections for headache treatment [21].

REFERENCES

[1] M. Appasamy, C. Lam, J. Alm, A.L. Chadwick, Trigger point injections, Phys. Med. Rehabil. Clin 33 (2) (2022) 307−333, https://doi.org/10.1016/j.pmr.2022.01.011.

[2] C. Bron, J.D. Dommerholt, Etiology of myofascial trigger points, Curr. Pain Headache Rep. 16 (5) (2012) 439−444, https://doi.org/10.1007/s11916-012-0289-4.

[3] J.P. Shah, T.M. Phillips, J.V. Danoff, L.H. Gerber, An in vivo microanalytical technique for measuring the local biochemical milieu of human skeletal muscle, J. Appl. Physiol. 99 (5) (2005) 1977−1984, https://doi.org/10.1152/japplphysiol.00419.2005.

[4] M.S. Robbins, Clinic-based procedures for headache, Continu. Lifelong Learn. Neurol. 27 (3) (2021) 732−745, https://doi.org/10.1212/CON.0000000000000959.

[5] E.P. Calandre, J. Hidalgo, J.M. García-Leiva, F. Rico-Villademoros, Trigger point evaluation in migraine patients: an indication of peripheral sensitization linked to migraine predisposition? Eur. J. Neurol. 13 (3) (2006) 244−249, https://doi.org/10.1111/j.1468-1331.2006.01181.x.

[6] C. Fernández-De-Las-Peñas, C. Alonso-Blanco, M.L. Cuadrado, R.D. Gerwin, J.A. Pareja, Myofascial trigger points and their relationship to headache clinical parameters in chronic tension-type headache, Headache 46 (8) (2006) 1264−1272, https://doi.org/10.1111/j.1526-4610.2006.00440.x.

[7] A. Blumenfeld, A. Ashkenazi, B. Grosberg, U. Napchan, S. Narouze, B. Nett, T. Depalma, B. Rosenthal, S. Tepper, R.B. Lipton, Patterns of use of peripheral nerve blocks and trigger point injections among headache practitioners in the USA: results of the American headache society interventional procedure survey (AHS-IPS), Headache 50 (6) (2010) 937−942, https://doi.org/10.1111/j.1526-4610.2010.01676.x.

[8] C. Alonso-Blanco, A.I. De-La-Llave-Rincón, C. Fernndez-De-Las-Peñas, Muscle trigger point therapy in tension-type headache, Expert Rev. Neurother. 12 (3) (2012) 315−322, https://doi.org/10.1586/ern.11.138.

[9] O. Karadas, H.L. Gül, L.E. Inan, Lidocaine injection of pericranial myofascial trigger points in the treatment of frequent episodic tension-type headache, J. Headache Pain 14 (1) (2013), https://doi.org/10.1186/1129-2377-14-44.

[10] Ö. Karadas, L.E. Inan, Ü.H. Ulas, Z. Odabasi, Efficacy of local lidocaine application on anxiety and depression and its curative effect on patients with chronic tension-type headache, Eur. Neurol. 70 (1−2) (2013) 95−101, https://doi.org/10.1159/000350619.

[11] R. Noseda, R. Burstein, Migraine pathophysiology: anatomy of the trigeminovascular pathway and associated neurological symptoms, cortical spreading depression, sensitization, and modulation of pain, Pain 154 (1) (2013) S44−S53S44, https://doi.org/10.1016/j.pain.2013.07.021.

[12] C. Fernández-de-las-Peñas, M.L. Cuadrado, J.A. Pareja, M. Trigger Points, Neck mobility and forward head posture in unilateral migraine, Cephalalgia 26 (9) (2006) 1061−1070, https://doi.org/10.1111/j.1468-2982.2006.01162.x.

[13] A. Ashkenazi, W.B. Young, The effects of greater occipital nerve block and trigger point injection on brush allodynia and pain in migraine, Headache 45 (4) (2005) 350−354, https://doi.org/10.1111/j.1526-4610.2005.05073.x.

[14] J.M. García-Leiva, J. Hidalgo, F. Rico-Villademoros, V. Moreno, E.P. Calandre, Effectiveness of ropivacaine trigger points inactivation in the prophylactic management of patients with severe migraine, Pain Med. 8 (1) (2007) 65–70, https://doi.org/10.1111/j.1526-4637.2007.00251.x.

[15] E.P. Calandre, J. Hidalgo, J.M. Garcia-Leiva, F. Rico-Villademoros, A. Delgado-Rodriguez, Myofascial trigger points in cluster headache patients: a case series, Head Face Med. 4 (1) (2008), https://doi.org/10.1186/1746-160X-4-32.

[16] B. Jaeger, Are "cervicogenic" headaches due to myofascial pain and cervical spine dysfunction? Cephalalgia 9 (3) (1989) 157–164, https://doi.org/10.1046/j.1468-2982.1989.0903157.x.

[17] Z. Mohammadi, Z. Shafizadegan, M.J. Tarrahi, N. Taheri, The effectiveness of sternocleidomastoid muscle dry needling in patients with cervicogenic headache, Adv. Biomed. Res. 10 (1) (2021) 10, https://doi.org/10.4103/abr.abr_138_20.

[18] J.K. Roth, R.S. Roth, J.R. Weintraub, D.G. Simons, Cervicogenic headache caused by myofascial trigger points in the sternocleidomastoid: a case report, Cephalalgia 27 (4) (2007) 375–380, https://doi.org/10.1111/j.1468-2982.2007.01296.x.

[19] A. Labastida-Ramírez, S. Benemei, M. Albanese, A. D'Amico, G. Grillo, O. Grosu, D.H. Ertem, J. Mecklenburg, E. Petrovna Fedorova, P. Řehulka, F. Schiano di Cola, J. Trigo Lopez, N. Vashchenko, A. MaassenVanDenBrink, P. Martelletti, Persistent post-traumatic headache: a migrainous loop or not? The clinical evidence, J. Headache Pain 21 (1) (2020), https://doi.org/10.1186/s10194-020-01122-5.

[20] E.P. Baron, N. Cherian, S.J. Tepper, Role of greater occipital nerve blocks and trigger point injections for patients with dizziness and headache, Neurol. 17 (6) (2011) 312–317, https://doi.org/10.1097/NRL.0b013e318234e966.

[21] M.S. Robbins, D. Kuruvilla, A. Blumenfeld, L. Charleston, M. Sorrell, C.E. Robertson, B.M. Grosberg, S.D. Bender, U. Napchan, A. Ashkenazi, Trigger point injections for headache disorders: expert consensus methodology and narrative review, Headache 54 (9) (2014) 1441–1459, https://doi.org/10.1111/head.12442.

[22] M. Castaldo, H.Y. Ge, A. Chiarotto, J.H. Villafane, L. Arendt-Nielsen, Myofascial trigger points in patients with whiplash-associated disorders and mechanical neck pain, Pain Med. 15 (5) (2014) 842–849, https://doi.org/10.1111/pme.12429.

[23] C. Hammi, J.D. Schroeder, B. Yeung, Trigger Point Injection, 2023, p. 2023.

[24] D.J. Alvarez, P.G. Rockwell, Trigger points: diagnosis and management, Am. Fam. Physician 65 (4) (2002) 653–660.

[25] A. Ashkenazi, A. Blumenfeld, U. Napchan, S. Narouze, B. Grosberg, R. Nett, T. DePalma, B. Rosenthal, S. Tepper, R.B. Lipton, Peripheral nerve blocks and trigger point injections in headache management - a systematic review and suggestions for future research, Headache 50 (6) (2010) 943–952, https://doi.org/10.1111/j.1526-4610.2010.01675.x.

[26] J. Cheng, S. Abdi, Complications of joint, tendon, and muscle injections, Tech. Reg. Anesth. Pain Manag. 11 (3) (2007) 141–147, https://doi.org/10.1053/j.trap.2007.05.006.

[27] E.O. Ahiskalioglu, H.A. Alici, A. Dostbil, M. Celik, A. Ahiskalioglu, M. Aksoy, Pneumothorax after trigger point injection: a case report and review of literature, J. Back Musculoskelet. Rehabil. 29 (4) (2016) 895–897, https://doi.org/10.3233/BMR-160666.

[28] N. Shafer, Pneumothorax following "trigger point" injection, JAMA, J. Am. Med. Assoc. 213 (7) (1970) 1193, https://doi.org/10.1001/jama.1970.03170330073017.

[29] S.-C. Cho, D.-R. Kwon, J.-W. Seong, Y. Kim, L. Özçakar, A pilot analysis on the efficacy of multiple trigger-point saline injections in chronic tension-type headache: a retrospective observational study, J. Clin. Med. 11 (18) (2022) 5428, https://doi.org/10.3390/jcm11185428.

Chapter | Five

Sphenopalatine ganglion blocks for headaches

Ahish Chitneni[1], Amit Aggarwal[2], Edward Pingenot[1], Alan David Kaye[1]

[1]LSU Health Shreveport (Formerly LSU Health Sciences Center Shreveport), Department of Anesthesiology, Shreveport, LA, United States; [2]Department of Anesthesiology and Perioperative Medicine, University of Texas Medical Branch, Galveston, TX, United States

INTRODUCTION

Underlying the middle nasal turbinate and suspended from the maxillary nerve within the sphenopalatine fossa, the largest collection of cell bodies inside of the calvarium and outside of the brain has been identified as an effective anatomical target for treating a variety of craniofacial pain syndromes [1,2]. Also known as the pterygopalatine or Meckel's ganglion, the sphenopalatine ganglion (SPG) is the most cephalad sympathetic pathway for which diagnostic and therapeutic blocks are commonly performed [3]. Though primarily composed of parasympathetic fibers, the SPG also contains sympathetic and sensory fibers.

Parasympathetic afferent fibers arise from the superior salivatory nucleus of CNVII, continue through the greater superficial petrosal nerve, and travel through the vidian (pterygoid) canal to become the vidian nerve and eventually reach the SPG. Postganglionic parasympathetic fibers innervate the mucosa of the nose, soft palate, uvula, tonsils, gums, upper lips, roof of the mouth, pharynx, lacrimal glands, and meningeal vessels.

Sympathetic fibers from superior cervical ganglion travel along the internal carotid plexus to become the deep petrosal nerve and pass through the vidian canal as the vidian nerve to reach the SPG. Sympathetic fibers supply nasal and pharyngeal mucosa as well as the lacrimal gland.

Sensory fibers of the SPG arise from the maxillary nerve, with a majority of efferent fibers passing directly through and exiting the ganglion through the

Interventional Management of Migraines and Other Headache Disorders. https://doi.org/10.1016/B978-0-443-23557-3.00005-X

greater and lesser palatine nerves without synapsing, although a small number of axons may synapse with the SPG itself. The greater palatine nerve supplies pain sensation to the bony palate, as well as to the gingival and buccal mucosa. The lesser palatine nerve supplies pain sensation to the uvula, tonsils, and soft palate [4].

All branches arising from the SPG carry these three sets of fibers to the areas where they terminate [5]. Thus, significant pain relief might be achieved when associated pain signal pathways are effectively blocked. At the time of this writing, more data regarding the overall effectiveness of sphenopalatine ganglion blocks is warranted. However, prevailing medical literature continues to support the utilization of sphenopalatine ganglion blocks to treat a variety of craniofacial syndromes, including headaches.

INDICATIONS

With its easily accessible approach through the nasal mucosa, the sphenopalatine ganglion (SPG) gives rise to many branches of the sensory, sympathetic and parasympathetic nervous systems [6]. With its proximity to numerous pain receptors, the SPG block continues to be extensively studied as an interventional treatment for facial pain and headaches uncontrolled with oral medications and other conventional methods. A study from 2017 showcases that use of the SPG block in playing a central role in treating trigeminal autonomic cephalalgia, a collection of primary headache disorders, such as cluster headaches and paroxysmal hemicrania, presenting as moderate to severe unilateral pain, distributed among the first branch of the trigeminal nerve [7]. Patients also complain of sympathetic and parasympathetic side effects including ptosis, miosis, and lacrimation, rhinorrhea, and nasal congestion respectively. Alongside its autonomic connections, the SPG has neural connections to the brainstem where cluster headaches and migraines originate, and to the meninges through the trigeminal nerve [16]. Evidence continues to arise for the use of the SPG blocks to treat adult acute migraine attacks, and has even been utilized for less common types of headaches such as postdural lumbar puncture headaches and intranasal contact point headaches [8,9]. A retrospective study from 2021 researched that pediatric populations also have a high incidence of persistent headaches and migraines as the most common cause of pain, primarily in prepubertal males over females. The incidence increases from 8% of 3 year olds to almost 60%–80% in 8- to 15-year-olds [8,9]. In summary, cluster headaches, trigeminal neuralgia, migraines, postdural lumbar puncture headaches and myofascial pain are a few types of conditions that can be treated through blockage of the SPG.

With its connections through the postganglionic parasympathetic fibers connected to the cerebral hemisphere and the upper cervical roots connected to the superior cervical sympathetic ganglion, the SPG block is theoretically implemented to relieve neuropathic pain from the head, face, neck, and thoracic region [8,9]. Major contraindications to the block include the use of anticoagulation agents such as warfarin, heparin, and factor 10a inhibitors, history of facial trauma, nasal canal stenosis or atresia, and infection [10]. Patients who suffer from chronic migraines typically trial through preventive treatments with goals of decreasing frequency of attacks, headache intensity and duration, and improving overall function. Some conventional approaches must take into account comorbidities. For example, patients who have any prior history of myocardial infarction have a contraindication to triptan use [11]. Those who fail conventional approaches and have frequent, refractory episodic chronic migraines with inadequate control can qualify for SPG blocks with proper patient consent and minimal contraindications. Given that patients are only minimally sedated for the procedure and can undergo stress and trauma for the procedure, patients with unstable cardiac arrhythmias are discouraged from undergoing the block. The SPG block has been theorized to produce vasodilatory effects, which in turn would have a beneficial impact on patients who suffered from ischemic strokes or who currently endure migraines without aura [11].

A bilateral transnasal SPG block has continued to be analyzed as a therapy for postdural puncture headaches (PDPH) in the obstetrical population. Prior to the SPG, healthcare professionals relied on an epidural block patch, which is injecting approximately ~20 mL of the patient's blood through an epidural needle into the epidural space. Major risk factors for PDPH include pregnancy, female gender, and young age, and can often lead to the complication of acquiring chronic headaches [12,13]. Given that research is still actively being conducted, only a limited patient population with PDPH can be selected to undergo an SPG block in emergency department settings and in the post anesthesia care units [8,9]. Some case reports have a wide variety of analgesia. The pain relief with 4% lidocaine from an SPG block lasted about 4–6 hours as a first line treatment for PDPH while other cases report up to 18 hours in those where epidural block patches are contraindicated [8,9]. Those patients who were contraindicated for epidural block patches were similar for those contraindicated for SPG blocks, which included coagulopathy and systemic infections. Unfortunately, current research also shows that the pain relief is temporary, and no statistical or clinical significance was found in patients who received 4% lidocaine or other local anesthetics compared with saline for pain intensity.

TECHNIQUE

As previously mentioned, various indications for the sphenopalatine ganglion block exist. Typically, the procedure is conducted through a transnasal approach as shown in Fig. 5.1 [8,9]. Initial positioning of the patient includes supine positioning as shown in Figure 5.10.5 mL of 0.5% bupivacaine is administered to each nostril while the swab is left in place for a total of 15 to 20 minutes while asking the patient to sniff to draw the anesthetic in Ref. [14]. After proper positioning with cervical extension, a sterile 10-cm cotton swab is dipped in 2% viscous lidocaine and advanced until reaching the posterior wall of the nasopharynx [14]. After 20 minutes, the cotton swab applicators are removed

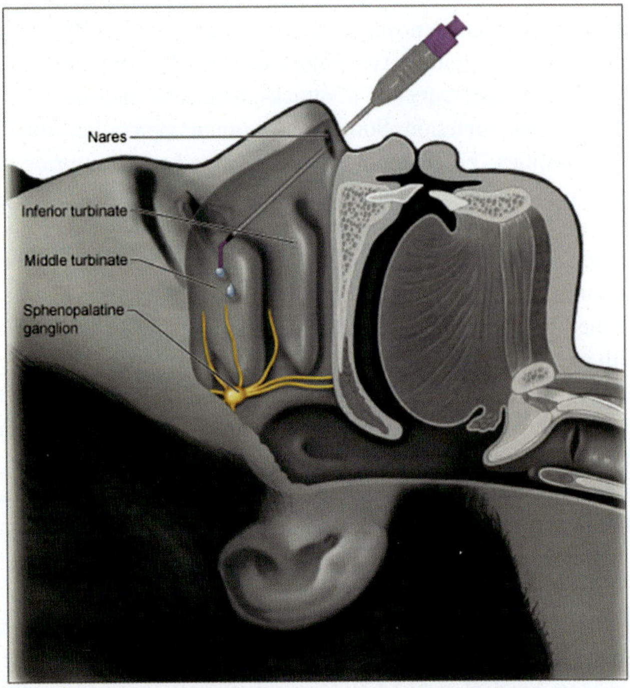

FIGURE 5.1 Intranasal sphenopalatine ganglion block technique [10].

from the nostril. During the procedure, blood pressure, heart rate, and respirations are monitored for any side effects [14]. During the procedure, various side effects and complications may occur that require monitoring. Orthostatic hypotension is a possible side effect of the procedure and can be monitored by patient symptoms and blood pressure monitoring. Epistaxis is another major complication that may occur and requires monitoring [8,9].

An alternative to the transnasal approach to a sphenopalatine ganglion block has been proposed in literature. Another approach includes a suprazygomatic, percutaneous, needle approach to approach the SPG for a block has been utilized under ultrasound guidance [15]. In this approach, the ultrasound probe is placed on the face under the zygomatic arch and angled 45 degrees cephalad to image the pterygopalatine fossa as shown in Fig. 5.2. After proper positioning of the probe, a needle is inserted superior to the zygomatic arch and advanced posteriorly to deliver 5 mL of 0.5% ropivacaine.

EFFICACY AND SAFETY

The efficacy of the SPG block has been widely studied, primarily for its use in cluster headaches and postdural puncture headaches. A literature analysis of multiple case reports/series and studies in 2017 focused on different agents used for the block. After induction with nitroglycerin to create a cluster

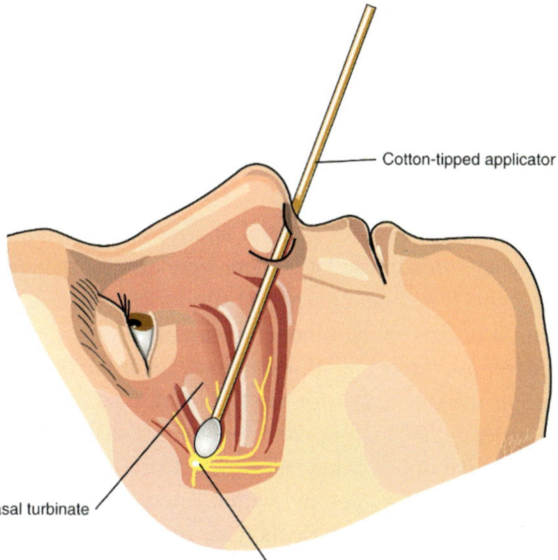

FIGURE 5.2 Application of local anesthetic via transnasal approach prior to procedure.

headache, patients were distributed evenly to be receive the following agents nasally near the sphenopalatine fossa for the SPG block: 1 mL of 10% cocaine hydrochloride, 1 mL of 10% lidocaine, or 1 mL of saline. It was found that saline produced the longest relief time compared with lidocaine and cocaine, with a 50%–100% relief. Due to the short-term relief, a mixture of steroid and other local anesthetics were added to improve the severity, consisting of triamcinolone, bupivacaine, mepivacaine, and epinephrine. It was shown to improve the severity and frequency of cluster headaches for half of the 21 patients in a different study [7]. A metaanalysis of many case reports, case series, and retrospective studies shows that the strongest evidence in using the SPG block is most useful for cluster headaches through injections, radiofrequency ablation, and neurostimulation [7].

Other evidence of SPG blocks shows significant relief for postdural puncture headaches in obstetrical populations. A majority of the studies focused on utilizing lidocaine only or saline only for the SPG blocks. A recent study from 2022 shows a combination of lidocaine and dexamethasone in each nostril showed prolonged duration of the block with faster onset, reporting less pain intensity. Application of the compounds were either administered through cotton-tipped swabs or a nasal applicator. Effectively treating PDPH can sometimes be complicated, as providers are taught to recommend rest, hydration, caffeine, and other analgesics. Epidural blood patches continue to be the gold standard for treating PDPH. However, due to its invasiveness, this procedure can lead to serious complications including neurological insult and multiple dural punctures. Utilization of the SPG block shows decreased need and incidence of epidural blood patches with its subsequent complications [16].

Analysis of sedation for the SPG block in pediatric patients was also extensively studied for the treatment of migraine headaches. In an extensive study that altered the administration of sedation, whether patients received intravenous versus oral midazolam or intravenous diphenhydramine. The study published findings that the pain reduction scale was not statistically significant in those who received sedation or general anesthesia compared with those who did not receive any sedative agents. There were also no statistically differences in gender, as females and males reported similar outcomes of pain relief. However, if sedation or general anesthetic is warranted for the noninvasive 10 minutes procedure, providers are allowed to act accordingly [16].

Some major disadvantages for the SPG block include major side effects, primarily transient mild to moderate loss of sensation within the maxillary nerve regions, in the upper gums and cheek. Only a certain percentage of patients have resolution of symptoms within 3 months when using an injection technique. Other major side effects can include bleeding, infection, and epistaxis [7]. Those who undergo radiofrequency thermal ablation for the SPG also report temporary postoperative epistaxis, cheek hematomas, and hypesthesia of the palate, with resolution within 3 months. Similar side effects are reported for patients who undergo neurostimulation of the ganglion [7].

Overall, numerous studies have shown the effectiveness and safety of the SPG block with satisfaction of multiple types of headaches, primarily migraines, cluster headaches, and PDPH. The side effects typically resolve within weeks to months with limited complications, and new projects continue to develop on what medications are ideal for the SPG block, whether local

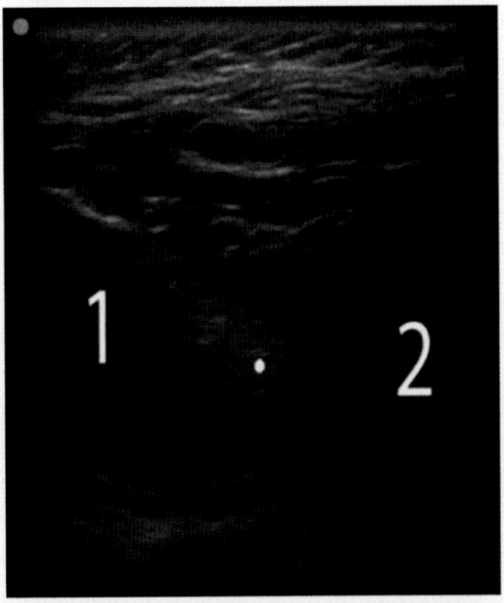

FIGURE 5.3 Ultrasound imaging showing (1) Maxilla and (2) Mandibular Ramus and pterygoid process.

anesthetics, saline, steroids, or combinations involving each of these classes. While there are numerous methods of controlling headaches, the SPG block has been shown to effectively treat headaches refractory from medications and other analgesic techniques. Proper discussion of the risks and benefits, with current research must be shared with each patient undergoing the SPG block, with limitations based on contraindications (Fig. 5.3).

REFERENCES

[1] M. Parviz Janfaza, J.B. Nadol Jr., J. Robert, M.S.I. Galla, L. Richard, M.D. Fabian, W.W. Montgomery, Surgical Anatomy of the Head and Neck, Vol First Harvard University Press edition. Harvard University Press, 2011, p. 2011.

[2] M.D. Nadya, Wilson P.E. Mitchell, 2019 Essentials of interventional cancer pain management springer sympathetic nervous system blocks for the treatment of cancer pain 17 (2019).

[3] R.J. Shah, B. Dixon, D. Padalia, Sphenopalatine Ganglion Radiofrequency Thermocoagulation, StatPearls Publishing, Treasure, 2022.

[4] N. Norton, Netter's Head and Neck Anatomy for Dentistry, Elsevier, 2017, pp. 257–274.

[5] K.W.D. Ho, R. Przkora, S. Kumar, Sphenopalatine ganglion: block, radiofrequency ablation and neurostimulation - a systematic review, J. Headache Pain 18 (1) (2017), https://doi.org/10.1186/s10194-017-0826-y.

[6] H.C. Diener, C. Tassorelli, D.W. Dodick, Management of trigeminal autonomic cephalalgias including chronic cluster: a review, JAMA Neurol. 80 (3) (2023) 308–319, https://doi.org/10.1001/jamaneurol.2022.4804.

[7] M.M. Mowafi, R.A. Abdelrazik, Efficacy and efficiency of sphenopalatine ganglion block for management of post-dural puncture headache in obstetric patients: a randomized clinical trial, Ain-Shams J. Anesthesiol. 14 (1) (2022) 73, https://doi.org/10.1186/s42077-022-00274-7.

[8] S.D. Waldman, Sphenopalatine ganglion block via the transnasal approach, in: Atlas of Pain Management Injection Techniques, fifth ed., Elsevier, Amsterdam, 2023, pp. 65–67.

[9] S.D. Waldman, Sphenopalatine Ganglion Block via the Transnasal Approach, Elsevier BV, 2023, pp. 65–67, https://doi.org/10.1016/b978-0-323-82826-0.00018-3.

[10] A. Forrest, A. Cantos, D. Butani, How we do it: sphenopalatine ganglion blockade for migraine treatment, Am. J. Interv. Radiol. 2 (14) (2018) 1–4.

[11] A.S. Nair, B.K. Rayani, Sphenopalatine ganglion block for relieving postdural puncture headache: technique and mechanism of action of block with a narrative review of efficacy, Kor. J. Pain 30 (2) (April 2017) 93–97, https://doi.org/10.3344/kjp.2017.30.2.93.

[12] Y. Jion, M.S. Robbins, The Sphenopalatine Ganglion (SPG) and Headache: AMF, American Migraine Foundation, September 10, 2018. https://americanmigrainefoundation.org/resource-library/sphenopalatine-ganglion/. (Accessed 28 June 2023).

[13] B.T. Bateman, N. Cole, C. Sun-Edelstein, C.L. Lay, Post Dural Puncture Headache, UpToDate, November 3, 2022. https://www.uptodate.com/contents/post-dural-puncture-headache?sectionName=Alternative+treatments+for+PDPH&topicRef=4469&anchor=H675550085&source=see_link#H675550085. (Accessed 28 June 2023).

[14] N. Padhy, S. Moningi, D.K. Kulkarni, R. Alugolu, S. Inturi, G. Ramachandran, Sphenopalatine ganglion block: intranasal transmucosal approach for anterior scalp blockade - a prospective randomized comparative study, J. Anaesthesiol. Clin. Pharmacol. 36 (2) (2020 Apr-Jun) 207–212, https://doi.org/10.4103/joacp.JOACP_249_18.

[15] M. Anthony Cometa, Y. Zasimovich, C.R. Smith, Percutaneous sphenopalatine ganglion block: an alternative to the transnasal approach, Int. J. Obstet. Anesth. 45 (February 2021) 163–164, https://doi.org/10.1016/j.ijoa.2020.10.002.

[16] M.A. Mousa, D.J. Aria, A.A. Mousa, C.M. Schaefer, M.H.H. Temkit, R.B. Towbin, Sphenopalatine ganglion nerve block for the treatment of migraine headaches in the pediatric population, Pain Physic. 24 (1) (January 2021) E111–E116. PMID: 33400444, https://www.painphysicianjournal.com/current/pdf?article=NzE5OQ%3D%3D.

FURTHER READING

[1] C.E. Alexander, A. Dua, Sphenopalatine ganglion block [Updated 2022 Nov 16], in: StatPearls, StatPearls Publishing, Treasure Island (FL), January 2023, https://www.ncbi.nlm.nih.gov/books/NBK557751/.

Chapter | Six

Cervical epidural steroid injections for the treatment of migraines and headaches

Hannah W. Haddad[1], Isaac Springer[2], Daniel Wang[3], Emily X. Zhang[4], Ivan Urits[5], Jamal J. Hasoon[6]

[1]Louisiana State University Health Science Center, Louisiana State University Physical Medicine and Rehabilitation Residency, New Orleans, LA, United States; [2]Oklahoma State University Medical Center, Oklahoma State University Anesthesiology Residency, Tulsa, OK, United States; [3]MedStar Georgetown University Hospital, Medstar Health Internal Medicine Residency, Washington, DC, United States; [4]Massachusetts General Hospital, Department of Anesthesiology, Critical Care, and Pain Medicine, Boston, MA, United States; [5]Southcoast Health, Brain and Spine, Wareham, MA, United States; [6]University of Texas Health Science Center, Department of Anesthesiology, Critical Care, and Pain Medicine, Houston, TX, United States

INTRODUCTION: CERVICAL ESI FOR MIGRAINES AND HEADACHES

Migraines affect millions of individuals worldwide daily. Because they are often debilitating, patients often have a significantly impacted quality of life and level of function. Treatments for headaches and migraines often start conservatively with lifestyle modifications and over-the-counter nonpharmaceutical and pharmaceutical management. If that does not provide adequate relief, more interventional treatments such as cervical epidural steroid injections (CESIs) have emerged as a noteworthy interventional therapy. This offers an additional form of treatment and relief for the patients to tackle the grappling burden of persistent migraines. This introduction will explore the role of CESIs and their use for migraines and headaches.

Interventional Management of Migraines and Other Headache Disorders. https://doi.org/10.1016/B978-0-443-23557-3.00006-1

Understanding cervical epidural steroid injections

CESIs are a minimally invasive treatment that deliver antiinflammatory mediators, typically corticosteroids, into the epidural space surrounding the cervical spinal cord. It is the outermost part of the spinal canal and hosts nerve roots and protective portions of the spinal cord. Introducing steroids to this area reduces inflammation and alleviates pain. This in turn improves overall spinal function and mobility [1].

Typically, this procedure is performed under fluoroscopy and X-ray guidance to ensure precision and efficacy. A needle is inserted into the cervical epidural space of the target level and medication is inserted targeting the source of pain and inflammation. CESIs can be used to treat conditions such as degenerative disc disease, spinal stenosis, disc herniations, and nerve root compression [2].

CESIs in the context of migraines and headaches

While CESIs are most used to target cervical spinal pathologies as listed above, they are increasingly being used for headache pathologies. Increasing evidence suggests certain types of headaches, specifically cervicogenic, may benefit from CESIs.

Cervicogenic headaches are characterized by pain originating in the neck/cervical spine area that radiates to the head. The neck is typically considered as the source given factors such as nerve irritation, joint dysfunction, and muscle tension all surround the neck region. As such, CESIs offer a specific targeted method of pain relief by targeting the underlying cervical spine issue [3]. Several studies, and increasingly more, have explored the efficacy of CESIs in headache management. These studies have shown promising results, which shows that CESIs may lead to significant pain reduction and increased quality of life for individuals with cervicogenic headaches [4].

CESIs show promising potential for managing migraines and headaches, specifically for cervical spine issues. These minimally invasive procedures offer pinpoint relief and have shown to improve quality of life for patients. As further research accumulates, CESIs are likely to play an increasingly important role in management of headaches. This chapter will provide further details.

MECHANISM OF ACTION OF THERAPEUTIC RELIEF

CESIs have gained recognition as an interventional modality for migraines and headaches, specifically cervicogenic headaches. Understanding the mechanism of action behind CESIs is critical to help practitioners appreciate the therapeutic efficacy and to create appropriate treatment plans for those suffering from these debilitating conditions.

Reduction of inflammation

One of the primary roles that CESIs play is the reduction of inflammation. This is important because inflammation plays a critical role in the pathophysiology of various headaches. CESIs utilize the antiinflammatory effects of corticosteroids to mitigate the inflammation involved with some specific spine pathologies. These steroids, which often include dexamethasone and/or methylprednisolone, inhibit the production of proinflammatory cytokines and enzymes. This ultimately leads to the desensitization of pain receptors and reduces edema [5]. By decreasing this inflammation within the cervical epidural space and nearby structures, CESIs address one of the primary underlying causes of headache pathologies.

Decreasing neural irritation

In the cervical epidural space, it is extremely narrow and small and houses a lot of structures such as spinal nerve roots. Because of the structures in such close proximity, the spinal nerve roots can often be irritated or compressed due to pathology such as spinal stenosis, degenerative changes, and disc herniations. If these nerves are irritated, they may transmit pain signals to the head, which may be interpreted as headaches. CESIs provide a therapeutic option by injecting corticosteroids and local anesthetics. The corticosteroids will aid in reducing nerve root inflammation and the anesthetics will provide fast-acting pain relief [5].

Muscle relaxation

Another etiology that contributes to cervicogenic headaches includes muscle spasm and tension. There are many causes of the muscle spasms but some of the most often include underlying spinal conditions and stress. CESIs can induce muscle relaxation around the cervical spine and thus indirectly improving headache symptoms. By reducing these muscle spasms and tension, CESIs aid in improving the pain that originates from the soft tissues and thus will improve neck mobility and headache relief [5].

Pain signal modulation

Many chronic pain conditions, including headaches, include modified pain signaling pathways. CESIs may aid in modulating these pathways. Steroids have been shown to influence the excitability and activation of nerve cells and the reduction of transmission of pain signals. They can also disrupt the cycle of chronic pain by interrupting the ongoing pain signal transmission from the cervical spine to the brain. This type of modulation helps break the cycle of headache pathologies and increases pain relief [6].

Antiedema

Swelling or edema in any region of the spine, specifically the cervical area where structures are very close in proximity, can exacerbate pain especially for cervicogenic headaches. ESIs help decrease the swelling within the spinal epidural space and surrounding structures [7,8]. Through reducing this edema, CESIs subsequently reduce the mechanical pressure on nerve roots and adjacent structures including spinal anatomy and blood vessels. This further contributes to increased headache relief [9].

Suppression of the immunological system

Some types of headaches have autoimmune and immune components. Thus, they may benefit from immunomodulatory effects of corticosteroids given that corticosteroids possess some immunosuppressive properties. As such, CESIs Hope regulate the immune response in the service of the spine region by providing a targeted delivery system for the steroids to utilize the immunosuppressive properties [9].

In summary, CESIs have many mechanisms of action both direct and indirect to provide therapeutic relief for migraines and headaches. They reduce inflammation, alleviate neural irritation, decrease muscle tension, change pain signaling, reduce edema, and suppress immune responses. These combined mechanisms provide an effective method of headache relief and improve quality of life for individuals suffering from headache pathologies.

SAFETY AND COMPLICATIONS OF CERVICAL EPIDURAL STEROID INJECTIONS

Epidural steroid injections of the cervical spine can be divided into Cervical Interlaminar Epidural Steroid Injections (CILESI) and Cervical Transforaminal Epidural Steroid Injections (CTFESI). Reporting of complications, as it relates to these two approaches, has been hindered by current coding systems not differentiating between approaches for epidural injections. One study reported that most adverse events have come from the transforaminal approach and prevention of these complications could be avoided given certain considerations. These considerations included, efficacy of treatment, medical necessity, and indications for treatment, pathoanatomy and mechanism of complications, alternate techniques, alternate injectable medications, and appropriate monitoring. This specific study also stated that CTFESI and Thoracic Transforaminal Epidural Steroid Injection (TTFESI) account for only 2.4% of all epidural steroid injections but account for 99% of all reported intraarterial injections of particulate steroids. This data, therefore, has been incorrectly applied to interlaminar epidural steroid injections of the caudal, lumbar, cervical, and thoracic spine [10]. It is also important to note that pain-related treatments are frequently conducted on older, frail, or overweight individuals who often have multiple coexisting medical conditions [11].

Neurological harm caused by CTFESI can result from the introduction of steroid particles into the bloodstream through intraarterial injection. This underscores the importance of using steroids that are devoid of particles [12]. Although CTFESI and particulate steroid injections have come under recent scrutiny from the medical community, as mentioned above, there has also been a defense of this approach. In a recent study there was recommendations to have clear, reproducible protocols for every CTFESI and this study indicated the efficacy/safety of particulate steroids due to their lower solubility coefficient, less wash out of medication, and thereby having positive outcomes on tissues in comparison to nonparticulate steroids [13].

Complications from CTFESI and CILESI can be divided into major and minor criteria for complications. Minor complications have been identified as accidental disc puncture, transient exacerbation of pain, accidental dural puncture, pain at injection, persistent numbness, and vasovagal response. Vasovagal risk is much higher in the cervical region. In a recent study, major complications have been defined as needle trauma, ophthalmic complications, and neurovascular injury [14]. The study reports that trauma to neurovascular structures is more often due to needle trauma as opposed to specific injectable medicines and the risk is mitigated by withdrawing the needle when a patient reports paresthesia. Furthermore, due to the need for direct patient feedback, deep sedation with these types of procedures should be avoided if at all possible. Major complications are rare given that there has been minimal reporting in the estimated nine million steroid injections each year. In the context of interlaminar injections, complications predominantly arise from injury to the arteries supplying the spinal cord, the neurotoxic effects of medications, the presence of solvents and preservatives in these medications, and, finally, embolization resulting in ischemia. Due to the risk of complications, both major and minor, it is recommended that these types of procedures be reserved for pain specialists with experience and that these injections be done with image guidance [14].

Complications associated with fluoroscopy guided CILESI include dural puncture headache, increased neck pain, stiffness, intracranial hypotension, and vasovagal reactions [15]. In the event a patient reports paresthesia under fluoroscopic guidance, the needle should be withdrawn and an adjacent cervical level should be chosen. Inadequate patient response secondary to deep sedation coupled with cervical disc herniation and limited epidural space resulted in spinal cord injuries in two case reports [15,16]. Certain steps can be taken to mitigate risk and increase safety. These steps include reviewing available/previous imaging studies such as CT/MRI, estimating dermal to epidural distance with different imaging modalities, limiting total volume of injectate, avoiding deep sedation, and staying as low on the cervical spine when possible due to increasing thickness of epidural space caudally [15]. However, the epidural space in the cervical spine can be difficult to assess secondary to the ligamentum flavum being unfused at midline in roughly 70% of patients; for this reason, loss of resistance (LOR) technique becomes unreliable. As one moves caudad on the cervical spine, the scapula, shoulder, and bony upper extremity obstruct epidural views. This can ultimately lead to decreased needle visualization and place patients at risk of complications [16]. Reducing the risk associated with CILESI involves employing the contralateral oblique technique, minimizing the use of sedation, conducting the procedure at the C6–C7 level or lower, maintaining a sterile approach, and adhering to anticoagulation protocols [12]. While adherence to anticoagulation protocols is crucial in reducing the risk of bleeding, a recent study has indicated a 0.4% complication rate (comprising a fatal stroke and a fatal myocardial infarction) among those who ceased anticoagulants/antiplatelet agents in line with ASRA guidelines [17].

COMPARISON OF TECHNIQUES OF CERVICAL EPIDURAL STEROID INJECTION FOR PAIN SYMPTOMS

Cervical transforaminal epidural steroid injections

During CTFESI, patients are positioned in supine, oblique, or lateral. An anteroposterior (AP) fluoroscopic view is taken to assess bony anatomy, while an oblique view is utilized to evaluate the trajectory of the neuroforamen. Lidocaine is administered to numb the skin and expected needle track posterior to the neuroforamen. Throughout the process, a 25-gauge spinal needle is guided using serial fluoroscopy, with the aim of directing it toward the anterior half of the superior articular process, precisely one needle width posterior to the neuroforamen. This path ensures the avoidance of vertebral and radiculomedullary arteries, as well as spinal nerve roots. Ensuring the needle contacts the periosteum of the superior articular process in the oblique view prevents it from being directed too medially toward the thecal sac. The needle's placement is evaluated in both the AP and oblique views, and its tangential advancement is utilized to pass the spinal needle anterior to the superior articular process by a small margin. Furthermore, its progression is monitored in the AP view until the needle tip aligns with the midline of the articular pillars, with caution exercised to restrict its posterior advancement beyond the uncinate process's medial boundary line, thereby averting any threat to the thecal space. Throughout the process, the needle's positioning in the neuroforamen is ascertained, and its medial depth is gauged using an AP fluoroscopic view. Care is taken to avoid any contact between the needle and the spinal nerves or ventral ramus, which could result in severe radicular pain or paresthesia. If such symptoms occur, the needle is promptly withdrawn, and the procedure is aborted if the pain persists. Given the need for direct patient feedback during the process, deep sedation is avoided. Prior to the final needle placement, both AP and oblique views are obtained to confirm that the needle is situated in the posterior aspect of the neuroforamen (oblique view) and not medially to the uncinate process (AP view). Aspiration is attempted; however, this does not guarantee that the needle is not intravascular, followed by the injection of contrast medium under fluoroscopy to confirm the spread surrounding the spinal nerve and the entrance into the lateral epidural space, indicating the correct needle position [18]. While patients may experience pain like their radicular symptoms, proceeding with the procedure is permissible if the pain is tolerable, as this is typically due to the pressure of the injection. Once the reassuring spread is verified and confirmed to be clear of arteries, veins, subdural spaces, or subarachnoid spaces, the safe injection of steroids can be administered [19].

Previous research has explored various techniques aiming to determine the appropriate needle angles for entering the neuroforamina during CTFESI. A prior study examined the orientation angles of the cervical neuroforamina, consequently inferring the needle entry angles for CTESIs conducted through the anterior lateral approach, with a focus on enhancing safety [20]. The findings of this study suggested a 95% confidence interval for the needle angle between 39.84 and 57.56 degrees [20]. It should be noted that this data serves as a reference point and may not represent the entire population, but it does offer a valuable estimate. A more recent study using axial MRI images, this study employed lines to denote essential anatomical landmarks for various approaches [21]. Measurement of angles on the axial sections of the MRI was conducted using Picture Archiving and Communication System (PACS) software. The first line was drawn from the midpoint of the two articular pillars and extended through the precise midline of the spinous process. In instances of a bifid spinous process, the line was directed through the midpoint of the bifid process. The second line was drawn parallel to the ventral lamina line, referred to as the conventional transforaminal approach line (CTAL). The third line was drawn parallel to the ventral margin at the midpoint of the SAP's ventral border, termed the new transforaminal approach line (NTAL). The NTAL angle, approximately 70 degrees, was identified as safer compared with the CTAL angle, which was approximately 50 degrees, particularly with regard to the risk of vascular injuries to vessels such as the vertebral artery (VA), internal carotid artery (ICA), and internal jugular vein (IJV). Nonetheless, further research is required to establish a consensus on the safe needle entry angle, considering the current lack of comprehensive research [21].

Cervical interlaminar epidural steroid injections

The patient is positioned in the prone posture, or alternatively, the seated or lateral decubitus position may be adopted when necessary. An AP view of the specific target area, typically C7-T1, is acquired with the intention of enhancing the visibility of the interlaminar space aperture diameter. While certain physicians favor the midline approach, lateral or paramedian approaches are also viable alternatives. Findings from research conducted in both the lumbar and cervical regions indicate the absence of a single superior method [19,22,23]. A 17 to 22 gauge needle is employed for this procedure. Under the guidance of serial fluoroscopy, the needle is gradually advanced, with regular assessments in the oblique and lateral views to ensure it does not penetrate too deeply upon contact with the lamina. Depending on the approach chosen, the needle is directed either cephalad toward the midline for the midline approach or directly cephalad for the paramedian approach. Ongoing imaging in the contralateral and oblique views serves to confirm the needle's precise placement during this stage of the procedure. The needle is slowly advanced until a LOR is felt at the epidural space. Final confirmation is obtained in the AP view, as well as either the lateral or oblique view, before the injectate is administered [19]. Several studies have endeavored to determine the optimal contralateral angles for observing CILESIs. One study highlights that during CILESIs, a contralateral oblique (CLO) view at 60° is more effective than other angles for visualizing the epidural space when the needle tip is positioned within the interlaminar space and falls within the margin of the spinous processes [24]. Conversely, when the needle tip is placed in the interlaminar space and positioned laterally to the spinous processes, a CLO view at 50° is deemed the most suitable [24].

EMERGING ULTRASOUND USE IN CERVICAL EPIDURAL STEROID INJECTIONS

Ultrasound has emerged as an invaluable tool for needle localization and administration of injectable medical therapies. However, in the setting of CESIs, fluoroscopy and CT guidance have been the standard of care. Recent studies have shown that ultrasound utilization has proven effective in identifying pertinent structures for CESI such as the ligamentum flavum, dura mater, and interspinous ligament. In addition, ultrasound application for needle tip localization in CESI has led to the theoretical risk reduction for dural puncture, post dural puncture headache, and spinal cord injury [25]. Ultrasound has also been employed for cervical medial branch blocks, superficial cervical plexus blocks, and CTFESI. One study showed the benefit of ultrasound guided CTFESI in patients that had pain refractory to ultrasound guided selective spinal nerve root (SNR) injections. These patients reported significant improvement 6 months post injection and this benefit was attributed to the transforaminal needle advancement under ultrasound guidance for further epidural spread that is not seen with typical SNR injections [26]. Furthermore, ultrasound proved noninferior to CT guidance for CTFESI and allows for the easy identification of local blood vessels to help minimize risks for vessel trauma [27]. Finally, other emerging uses of ultrasound are for the amount of medication spread during epidural injections, prepuncture diagnostics (such as epidural space depth), and skin to dural distance [28,29]. CESI are utilized in large volumes throughout the country for acute, chronic, and malignant pain management strategies for areas involving the neck, head, face, and upper extremities [28]. Ultrasound will continue to emerge as an effective diagnostic tool for injectable medical therapies providing valuable information to the provider for the benefit and safety of the patient. In doing so, it will also diminish the risks associated with current imaging modalities such as CT and fluoroscopy [25–29].

CLINICAL STUDIES: SAFETY AND EFFICACY

Overview

While the use of CESIs is a common treatment modality for cervical radiculopathy, and cervical intraarticular injections have been utilized to manage cervicogenic headaches (CGH), there is a notable gap in the existing literature concerning the application of CESI for migraines and headaches [19,30]. Five clinical studies, comprising of three retrospective cohort studies and two clinical trials, were utilized to examine the clinical efficacy of CESI for migraines and headaches [4,31–34].

Efficacy

Use of CESI for migraines and headaches has been shown to demonstrate short-term efficacy [31,33]. Martelletti et al. conducted a nonrandomized clinical trial with nine patients suffering from chronic cervicogenic headaches. A sharp decrease in pain intensity was noted within the first 12 hours postinjection, with effects persisting throughout the 4-week follow-up period. During the six-month follow-up however, no significant differences in pain scores were noted compared with baseline [31].

Two clinical studies have provided evidence of pain relief from CESI lasting up to 3 months postinjection [32,33]. Alvin et al. performed a retrospective cohort study with 50 patients who received CESI, reporting decreased pain scores and an improvement in quality-adjusted life years compared with baseline at the three-month follow-up [32]. Cronen et al. executed an uncontrolled clinical trial with 48 patients, noting that CESI relieved symptoms of tension type headaches for up to 3 months. Patients underwent a series of cervical steroid epidural nerve blocks until complete pain relief was achieved, with decreased pain scores reported at both the six-week and three-month follow-up [33]. Both studies did not provide follow-up data for longer than a three-month follow-up period [32,33].

The efficacy of CESI for pain relief over a six-month duration has also been documented [34]. He et al. conducted a retrospective cohort study with 37 patients, in which they investigated the use of a continuous epidural blockade to alleviate CGH symptoms. Patients reported reduced days with mild-moderate pain, decreased occurrence of severe pain, and reduced daily medication usage for up to 6 months postinjection. However, there was no significant difference in pain scores at the 12-month follow-up compared with baseline [34].

The only clinical study examined that offered long-term evidence of pain relief from CESI was conducted by Li et al. [4]. Patients with chronic CGH received a single CESI at the C2 level, and at the two-year follow-up, patients reported a reduction in pain scores compared with baseline. The Kaplan–Meier curve revealed that the median pain relief in the CESI group was 4 months, meaning that 50% of patients experienced pain relief for greater than 4 months [4].

Safety

Beyond transient side effects, no complications were documented in the literature [4,31–34]. Martelletti et al. reported three patients experiencing a flushing sensation in the face, lasting from 45 minutes to four hours postinjection [31]. Li et al. noted that some patients, including those without diabetes, were administered insulin due to reports of elevated blood sugar postintervention [4]. No studies reported serious adverse side effects resulting from the use of CESI for migraines and headaches. However, it is worth nothing that complications following CESI for other purposes have been documented [35]. A total complication rate of 16.8% was reported following interlaminar CESI, with minor events such as increased neck pain, transient headaches, vasovagal reactions, and facial flushing appearing to be the most common side effects [36]. Although relatively rare, instances of serious adverse events following CESI, such as epidural hematoma, spinal cord injury, cardiopulmonary arrest, and death, have also been documented [37–40]. With any medical intervention there are inherent risks, but due to the minimal documented adverse effects, the use of CESI for migraines and headaches appears to be a safe treatment modality [4,31–34].

To date, there remains a shortage of clinical studies examining the efficacy of CESI for migraines and headaches. The studies reviewed offer preliminary evidence that CESI can provide both short-term and long-term pain relief, with reported analgesia ranging from one hour to 2 years postinjection [4,31]. It remains unclear whether administering multiple CESI is more effective for long-term pain relief, as the evidence examined does not establish a clear correlation between the number of injections and the duration of pain relief [4,32–34]. Further investigations employing a randomized design and larger sample sizes are warranted in the future to determine the optimal therapeutic course of action of CESI for migraines and headaches.

CONCLUSION

Throughout this chapter, it has been demonstrated that CESIs have noteworthy efficacy and potential in the treatment of migraines and headaches. It demonstrated the pressing need for effective intervention in the realm of migraines and headaches given that they are extremely debilitating. Given that cervical epidural start injections are minimally invasive, they have promising value as a therapeutic option. Through its mechanism of action of reducing inflammation, alleviating early irritation, relaxing cervical muscles, ability to modulate pain signals, decreasing swelling, and suppressing the negative effects of the immune system, CESIs offer a multifaceted approach to migraine and headache management. Safety is also extremely important for any interventional procedure and CESIs are no exception. CESIs are generally safe, but it is also essential to be aware of potential risks to allow for informed decision making. Ultrasound is also an increasingly relevant form of imaging that shows promising potential in aiding to make CESIs more efficient and safe. In conclusion, CESIs represent a promising multifaceted approach to alleviating migraines and headaches. As further research is conducted, CESIs may emerge as a standard of care after failed conservative therapy in the comprehensive treatment of these conditions (Table 6.1).

Table 6.1 Summary of clinical studies on CESI efficacy.

Author and year	Study design	Participants	Indication(s)	Intervention(s)	Outcome measure(s)	Results	Duration of improvement	Adverse effects	Conclusion(s)
Martelletti et al. [31]	Nonrandomized clinical trial	9 patients	Diagnosis of cervicogenic headache, confirmed with GON or C blockades	CESI	Quality (none, moderate, worst possible), intensity of pain through the NIS, & DCI	Notable decrease in NIS scores during the first 12 hours postinjection. Both NIS and DCI indices remained significantly lowered compared with the control group during the 4 week period	1 hour - 4 weeks	3 patients reported a facial flushing sensation lasting 45 minutes to 4 hours postinjection	The cervical epidural steroid blockade proved efficient at decreasing pain characteristics and drug consumption
Alvin et al. [32]	Retrospective cohort study	50 patients	Neck pain and cervical radiculopathy for <6 months	At least one interlaminar or transforaminal epidural steroid injection (average 1.4 injections)	QOL scores using the EQ-5D, PDQ, and the PHQ-9	Significant reduction of the PDQ score compared with controls ($P = .05$), and clinically relevant improvement in EQ-5D score	3 months	None	Use of CESI's was efficient at improving quality-adjusted life years compared with the controls
Cronen et al. [33]	Uncontrolled clinical trial	48 patients	Intractable tension type headaches with no improvement following 1 month of traditional therapy, negative CT or MRI scan within 1 year	Cervical steroid epidural nerve blocks, administered every other day until complete pain relief was achieved (average 4 injections)	Pain VAS	Decreased mean VAS score at 6 weeks (0.95) and 3 months (0.35) compared with baseline mean VAS scores (4.8)	3 months	None	Use of cervical steroid epidural nerve blocks provided safe and effective relief of tension type headaches up to 3 months

Continued

Table 6.1 Summary of clinical studies on CESI efficacy. continued

Author and year	Study design	Participants	Indication(s)	Intervention(s)	Outcome measure(s)	Results	Duration of improvement	Adverse effects	Conclusion(s)
He et al. [34]	Retrospective cohort study	37 patients	Diagnosis of cervicogenic headache, initial VAS score of >60 mm, degeneration of cervical vertebra, and positive response to blocking C2-3	Continuous epidural block of a local anesthetic and corticosteroids, lasting 3–4 weeks	5 point scale, 0 (no pain)-5 (worst pain ever experienced)	Decreased days with mild-moderate pain, decreased occurrence of severe pain, and decreased daily NSAID use in 1–6 months postepidural	1–6 months	None	Continuous epidural blocks are effective at controlling chronic cervicogenic headaches for at least 6 months
Li et al. [4]	Retrospective cohort study	52 patients	Diagnosis of cervicogenic headache for >1 month, with no improvement following medical therapy	Cervical epidural injection of local anesthetic and steroid	Izbicki pain score including VAS, frequency of panic attacks, analgesic medication usage, and inability to work, NDI.	Significant reduction of total pain scores from before therapy (72.5) to the 2-year follow-up (40.00)	2 years	Some patients reported elevated blood glucose and required insulin to recover their blood sugar	Sustained pain relief was achieved for up to 2 years with CESIs

Cervical Epidural Steroid Injection (CESI); Drug disability index (DCI); EuroQol-5 Dimensions (EQ-5D); Neck disability index (NDI) ; Numeric Intensity Scale (NIS); Pain disability questionnaire (PDQ); Pain health questionnaire (PHQ-9); Quality of Life (QOL); Visual analogue scale (VAS).

REFERENCES

[1] K. Patel, P. Chopra, S. Upadhyayula, Epidural steroid injections, in: StatPearls, StatPearls Publishing, Treasure Island (FL), July 3, 2023.

[2] C. Ekhator, A. Urbi, B.N. Nduma, S. Ambe, E. Fonkem, Safety and efficacy of radiofrequency ablation and epidural steroid injection for management of cervicogenic headaches and neck pain: meta-analysis and literature review, Cureus 15 (2) (2023) e34932, https://doi.org/10.7759/cureus.34932.

[3] S. Diwan, L. Manchikanti, R.M. Benyamin, et al., Effectiveness of cervical epidural injections in the management of chronic neck and upper extremity pain, Pain Phys. 15 (4) (2012) E405–E434.

[4] S.J. Li, D. Feng, Pulsed radiofrequency of the C2 dorsal root ganglion and epidural steroid injections for cervicogenic headache, Neurol. Sci. 40 (6) (2019) 1173–1181, https://doi.org/10.1007/s10072-019-03782-x.

[5] R.F. McLain, L. Kapural, N.A. Mekhail, Epidural steroid therapy for back and leg pain: mechanisms of action and efficacy, Spine J. 5 (2) (2005) 191–201, https://doi.org/10.1016/j.spinee.2004.10.046.

[6] J.F. Howe, J.D. Loeser, W.H. Calvin, Mechanosensitivity of dorsal root ganglia and chronically injured axons: a physiological basis for the radicular pain of nerve root compression, Pain 3 (1) (1977) 25–41, https://doi.org/10.1016/0304-3959(77)90033-1.

[7] A. Berg, Clinical and myelographic studies of conservatively treated cases of lumbar intervertebral disk protrusion, Acta Chir. Scand. 104 (2–3) (1952) 124–129.

[8] L.N. Green, Dexamethasone in the management of symptoms due to herniated lumbar disc, J. Neurol. Neurosurg. Psychiatr. 38 (12) (1975) 1211–1217, https://doi.org/10.1136/jnnp.38.12.1211.

[9] E. Wang, D. Wang, Treatment of cervicogenic headache with cervical epidural steroid injection, Curr. Pain Headache Rep. 18 (9) (2014) 442, https://doi.org/10.1007/s11916-014-0442-3.

[10] L. Manchikanti, R.M. Benyamin, Key safety considerations when administering epidural steroid injections, Pain Manag. 5 (4) (2015) 261–272, https://doi.org/10.2217/pmt.15.17.

[11] G. Lo Bianco, A. Tinnirello, A. Papa, et al., Interventional pain procedures: a narrative review focusing on safety and complications. Part 1 injections for spinal pain, J. Pain Res. 16 (2023) 1637–1646, https://doi.org/10.2147/JPR.S402798.

[12] B.J. Schneider, S. Maybin, E. Sturos, Safety and complications of cervical epidural steroid injections, Phys. Med. Rehabil. Clin 29 (1) (2018) 155–169, https://doi.org/10.1016/j.pmr.2017.08.012.

[13] K. Bush, R. Mandegaran, E. Robinson, A. Zavareh, The safety and efficiency of performing cervical transforaminal epidural steroid injections under fluoroscopic control on an ambulatory/outpatient basis, Eur. Spine J. 29 (5) (2020) 994–1000, https://doi.org/10.1007/s00586-019-06147-2.

[14] K. Van Boxem, M. Rijsdijk, G. Hans, et al., Safe use of epidural corticosteroid injections: recommendations of the WIP benelux work group, Pain Pract. 19 (1) (2019) 61–92, https://doi.org/10.1111/papr.12709.

[15] A. Chang, D. Wang, Complications of fluoroscopically guided cervical interlaminar epidural steroid injections, Curr. Pain Headache Rep. 24 (10) (2020) 63, https://doi.org/10.1007/s11916-020-00897-1.

[16] Y. Perper, On the spinal cord injury during attempted cervical interlaminar epidural injection of steroids, Pain Med. 20 (4) (2019) 854–855, https://doi.org/10.1093/pm/pny173.

[17] S. Endres, A. Shufelt, N. Bogduk, The risks of continuing or discontinuing anticoagulants for patients undergoing common interventional pain procedures, Pain Med. 18 (3) (2017) 403–409, https://doi.org/10.1093/pm/pnw108.

[18] W.J. Sullivan, S.E. Willick, W. Chira-Adisai, et al., Incidence of intravascular uptake in lumbar spinal injection procedures, Spine 25 (4) (2000) 481–486, https://doi.org/10.1097/00007632-200002150-00015.
[19] L.M. House, K. Barrette, R. Mattie, Z.L. McCormick, Cervical epidural steroid injection: techniques and evidence, Phys. Med. Rehabil. Clin. 29 (1) (2018) 1–17, https://doi.org/10.1016/j.pmr.2017.08.001.
[20] B. Chen, L. Rispoli, T.P. Stitik, P.M. Foye, J.S. Georgy, Optimal needle entry angle for cervical transforaminal epidural injections, Pain Phys. 17 (2) (2014) 139–144.
[21] M.H. Karm, J.Y. Park, D.H. Kim, et al., New optimal needle entry angle for cervical transforaminal epidural steroid injections: a retrospective study, Int. J. Med. Sci. 14 (4) (2017) 376–381, https://doi.org/10.7150/ijms.17112.
[22] R. Gupta, S. Singh, S. Kaur, K. Singh, K. Aujla, Correlation between epidurographic contrast flow patterns and clinical effectiveness in chronic lumbar discogenic radicular pain treated with epidural steroid injections via different approaches, Kor. J. Pain 27 (4) (2014) 353–359, https://doi.org/10.3344/kjp.2014.27.4.353.
[23] J.Y. Yoon, J.W. Kwon, Y.C. Yoon, J. Lee, Cervical interlaminar epidural steroid injection for unilateral cervical radiculopathy: comparison of midline and paramedian approaches for efficacy, Kor. J. Radiol. 16 (3) (2015) 604–612, https://doi.org/10.3348/kjr.2015.16.3.604.
[24] J.Y. Park, M.H. Karm, D.H. Kim, J.Y. Lee, H.J. Yun, J.H. Suh, Optimal angle of contralateral oblique view in cervical interlaminar epidural injection depending on the needle tip position, Pain Physic. 20 (1) (2017) E169–E175.
[25] N. Maeda, M. Maeda, Y. Tanaka, Direct visualization of cervical interlaminar epidural injections using sonography, Tomography 8 (4) (2022) 1869–1880, https://doi.org/10.3390/tomography8040157.
[26] X. Zhang, H. Shi, J. Zhou, et al., The effectiveness of ultrasound-guided cervical transforaminal epidural steroid injections in cervical radiculopathy: a prospective pilot study, J. Pain Res. 12 (2018) 171–177, https://doi.org/10.2147/JPR.S181915.
[27] L. Yue, S. Zheng, L. Hua, et al., Ultrasound-guided versus computed tomography fluoroscopy-assisted cervical transforaminal steroid injection for the treatment of radicular pain in the lower cervical spine: a randomized single-blind controlled noninferiority study, Clin. J. Pain 39 (2) (2023) 68–75, https://doi.org/10.1097/AJP.0000000000001091.
[28] S.H. Kim, K.H. Lee, K.B. Yoon, W.Y. Park, D.M. Yoon, Sonographic estimation of needle depth for cervical epidural blocks, Anesth. Analg. 106 (5) (2008), https://doi.org/10.1213/ane.0b013e318168b6a8.
[29] S.W. Yoo, M.J. Ki, A.R. Doo, C.J. Woo, Y.S. Kim, J.S. Son, Prediction of successful caudal epidural injection using color Doppler ultrasonography in the paramedian sagittal oblique view of the lumbosacral spine, Kor. J. Pain 34 (3) (2021) 339–345, https://doi.org/10.3344/kjp.2021.34.3.339.
[30] M. Appeadu, N. Miranda-Cantellops, B. Mays, et al., The effectiveness of intraarticular cervical facet steroid injections in the treatment of cervicogenic headache: systematic review and meta-analysis, Pain Phys. 25 (6) (2022) 459–470.
[31] P. Martelletti, F. Di Sabato, M. Granata, et al., Epidural corticosteroid blockade in cervicogenic headache, Eur. Rev. Med. Pharmacol. Sci. 2 (1) (1998) 31–36.
[32] M.D. Alvin, V. Mehta, H Al Halabi, D. Lubelski, E.C. Benzel, T.E. Mroz, Cost-Effectiveness of cervical epidural steroid injections: a 3-month pilot study, Global Spine J. 9 (2) (2019) 143–149, https://doi.org/10.1177/2192568218764913.
[33] M.C. Cronen, S.D. Waldman, Cervical steroid epidural nerve blocks in the palliation of pain secondary to intractable tension-type headaches, J. Pain Symptom Manag. 5 (6) (1990) 379–381, https://doi.org/10.1016/0885-3924(90)90034-H.
[34] M wei He, N.J. xiang, Y na Guo, Q. Wang, L qiang Yang, J jie Liu, Continuous epidural block of the cervical vertebrae for cervicogenic headache, Chin. Med. J. 122 (4) (2009) 427–430.

[35] K.D. Candido, N."N". Knezevic, Cervical epidural steroid injections for the treatment of cervical spinal (neck) pain, Curr. Pain Headache Rep. 17 (2) (2013) 314, https://doi.org/10.1007/s11916-012-0314-7.

[36] K.P. Botwin, R. Castellanos, S. Rao, et al., Complications of fluoroscopically guided interlaminar cervical epidural injections, Arch. Phys. Med. Rehabil. 84 (5) (2003) 627–633, https://doi.org/10.1016/s0003-9993(02)04862-1.

[37] K.N. Williams, A. Jackowski, P.J.D. Evans, Epidural haematoma requiring surgical decompression following repeated cervical epidural steroid injections for chronic pain, Pain 42 (2) (1990) 197–199, https://doi.org/10.1016/0304-3959(90)91162-C.

[38] R. Baker, P. Dreyfuss, S. Mercer, N. Bogduk, Cervical transforaminal injection of corticosteroids into a radicular artery: a possible mechanism for spinal cord injury, Pain 103 (1–2) (2003) 211–215, https://doi.org/10.1016/s0304-3959(02)00343-3.

[39] B. Stauber, L. Ma, R. Nazari, Cardiopulmonary arrest following cervical epidural injection, Pain Physic. 15 (2) (2012) 147–152.

[40] L. Rozin, R. Rozin, S.A. Koehler, et al., Death during transforaminal epidural steroid nerve root block (C7) due to perforation of the left vertebral artery, Am. J. Forensic Med. Pathol. 24 (4) (2003) 351–355, https://doi.org/10.1097/01.paf.0000097790.45455.45.

Chapter | Seven

Cervical medial branch blocks for the treatment of cervicogenic headaches

Hannah W. Haddad[1], Daniel Wang[2], Changho Yi[3], Crystal Li[4], Ivan Urits[5], Jamal J. Hasoon[6]

[1]*Louisiana State University Health Science Center, Louisiana State University Physical Medicine and Rehabilitation Residency, New Orleans, LA, United States;* [2]*MedStar Georgetown University Hospital, Medstar Health Internal Medicine Residency, Washington, DC, United States;* [3]*McAllen Medical Center, University of Texas Rio Grande Valley Family Medicine Residency, Brownsville, TX, United States;* [4]*University of Maryland, School of Medicine, Baltimore, MD, United States;* [5]*Southcoast Physician Group, Pain Management, Wareham, MA, United States;* [6]*University of Texas Health Science Center, Department of Anesthesiology, Critical Care, and Pain Medicine, Houston, TX, United States*

INTRODUCTION: CERVICAL MEDIAL BRANCH BLOCK FOR THE TREATMENT OF CERVICOGENIC HEADACHE

Cervicogenic headache stands out among various headache types due to its association with cervical pathologies [1]. Its prevalence has been reported to range from 2.5% in the general population to 15%–20% among patients suffering from chronic headaches [2]. While cervicogenic headache is often characterized by a reduced range of motion in the cervical spine, exacerbated pain upon neck movement, and diffuse arm and shoulder discomfort, these distinctive features are not always present [3]. The presence of myofascial tenderness in the cervical spine or neck pain is a common occurrence in primary headaches, making the clinical diagnosis of cervicogenic headache challenging [4]. Moreover, there is ongoing debate concerning the optimal diagnostic criteria for cervicogenic headache [5]. Consequently, both the International Headache Society (IHS) and the Cervicogenic Headache International Study Group (CHISG) advocate diagnostic criteria that encompass local anesthetic

Interventional Management of Migraines and Other Headache Disorders. https://doi.org/10.1016/B978-0-443-23557-3.00007-3

Table 7.1 Diagnostic criteria of cervicogenic headache.

[redacted]		
11.2.1 Cervicogenic headache **A.** Any headache fulfilling criterion C **B.** Clinical and/or imaging evidence of a disorder or lesion within the cervical spine or soft tissues of the neck, known to be able to cause headache **C.** Evidence of causation demonstrated by at least two of the following: **1.** Headache has developed in temporal relation to the onset of the cervical disorder or appearance of the lesion **2.** Headache has significantly improved or resolved in parallel with improvement in or resolution of the cervical disorder or lesion **3.** cervical range of motion is reduced and headache is made significantly worse by provocative maneuvers **4.** Headache is abolished following diagnostic blockade of a cervical structure or its nerve supply **D.** Not better accounted for by another ICHD-3 diagnosis.	Major criteria	I. Symptoms and signs of neck involvement Ia. Precipitation of head pain, similar to the usually occurring one: (Ia1) by neck movement and/or sustained, awkward head positioning, and/or: (Ia2) by external pressure over the upper cervical or occipital region on the symptomatic side. Ib. Restriction of range of motion (ROM) in the neck. Ic. Ipsilateral neck, shoulder or arm pain of a rather vague, non-radicular nature, or—occasionally—arm pain of a radicular nature. II. Confirmatory evidence by diagnostic anesthetic blockades. III. Unilaterality of the head pain, without side shift.
	Head pain characteristics	IV. Moderate—severe, non-throbbing pain, usually starting in the neck. Episodes of varying duration, or: fluctuating, continuous pain.
	Other characteristics of some importance	V. Only marginal effect or lack of effect of indomethacin. Only marginal effect or lack of effect of ergotamine and sumatriptan. Female sex. Not infrequent occurrence of head or indirect neck trauma by history, usually of more than only medium severity.
	Other features of lesser importance	VI. Various attack-related phenomena, only occasionally present, and/or moderately expressed when present: (a) Nausea, (b) phono- and photophobia, (c) dizziness, (d) ipsilateral "blurred vision", (e) difficulties swallowing, (f) ipsilateral edema, mostly in the periocular area.

blocks (as outlined in Table 7.1) [6,7]. Though not obligatory in either set of criteria, the administration of local anesthetic injections to facet joints or cervical medial branch blocks (MBBs) is considered essential for clinicians in reaching a diagnosis [8].

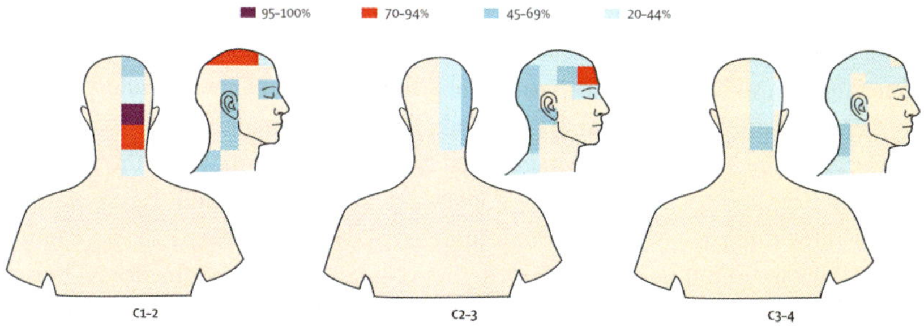

FIGURE 7.1 Areas of pain relief in patients who underwent controlled blocks of the synovial joints at C1–2, C2–3, and C3–4. The density of shading is proportional to the number of patients who perceived pain in the area indicated. Adapted from Cooper and colleagues, with permission from Blackwell Science.

When cervicogenic headache is highly suspected, the next step involves identifying the source of the pain generator [9]. History and physical examinations often produce nonspecific findings when it comes to identifying the exact source, but the pattern of pain distribution can offer hints regarding the origin of the pain generator (Fig. 7.1) [10]. Advanced imaging techniques, such as magnetic resonance imaging (MRI) scans, often fail to provide adequate diagnostic information [11]. The upper cervical spine harbors numerous pain-sensitive structures, including the facet joints, intervertebral discs, ligaments, and muscles. Among these, the facet joint is widely recognized as the most common pain generator, with the C2-3 joint being the primary culprit (accounting for 62% of cases), followed by the C1-2 (7%) and C3-4 (6%) joints [12,13]. Nevertheless, there have been reports of cervicogenic headache diagnoses and successful resolution through treatment targeting middle to lower facet joints [14–16]. In terms of prognostic indicators for radiofrequency ablation (RFA), MBBs offer greater predictive value than intraarticular facet injections [17]. To mitigate the risk of false-positive and false-negative results, a second confirmatory block is often recommended [9].

Once the diagnosis is confirmed, initial management typically involves physical therapy; medication has not been shown effective for cervicogenic headache [18,19]. Chiropractic manipulation has been studied and shown to provide short-term benefits [20–23]. In cases of refractory or long-term pain relief, interventional procedures or surgery may be considered. RFA is a commonly employed treatment method that has been used for Atlantoaxial joint, C2 dorsal root ganglion, third occipital nerve and mid-cervical medial branches [16,24–30]. Due to the potential risks associated with neurovascular injury to spinal nerve roots and vertebral arteries, mid-cervical branch blocks are often preferred and have reported successful outcomes [14].

Since RFA and MBBs share similar procedural techniques, a comprehensive understanding of the mechanism of action, potential pitfalls, and technical aspects of cervical medial branches is of paramount importance in both diagnosing and treating cervicogenic headaches.

MECHANISM OF ACTION OF THERAPEUTIC RELIEF

Understanding the pathophysiology of cervicogenic headaches originating from facet joints is pivotal for comprehending the mechanisms underpinning the therapeutic efficacy of MBBs in headache management.

Cervicogenic headache, characterized by pain emanating from the cervical spine, predominantly hinges on the convergence of cervical and trigeminal afferents within the trigeminocervical nucleus [8,31,32]. Situated in the caudal part of the trigeminal nucleus, this structure courses along with the dorsal horns of the upper three or four segments of the cervical spinal cord (Fig. 7.2). The trigeminocervical nucleus receives nociceptive inputs from spinal nerves C1, C2, and C3, which converge onto second-order neurons that also receive nociceptive signals from other cervical nerves and the V1 branch of the trigeminal nerve, including dural and occipital afferents [31–34]. This convergence provokes headaches in various regions, primarily the occipital and auricular areas, while also extending to parietal, frontal, and orbital regions due to the involvement of trigeminal afferents [35,36]. Similarly, nociceptive signals from the neck, triggering neck pain, can manifest as diverse headache types due to irritation of the dura [31,32]. Amid the various structures that may elicit pain in the cervical region, C2/3 and C3/4 facet joints have garnered substantial attention as pain generators, with the atlanto-axial (AA) and atlanto-occipital (AO) joints also emerging as potential sources of cervicogenic headaches [10,13,37,38].

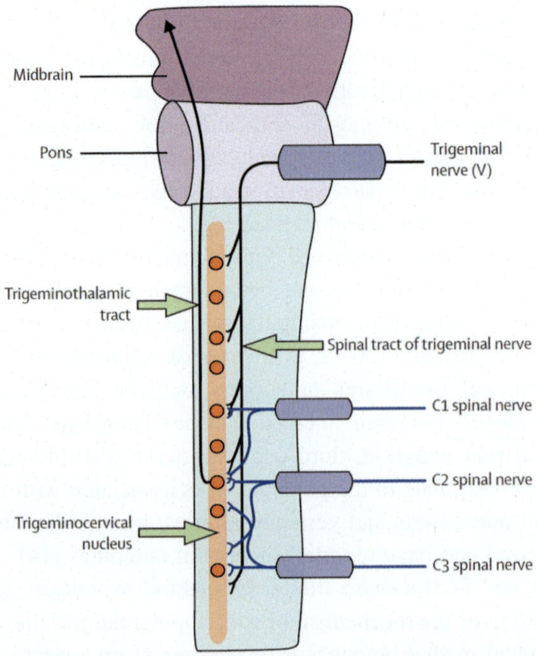

FIGURE 7.2 Schematic diagram of convergence in trigeminocervical nucleus. Derived from: Bogduk N, Govind J. Cervicogenic headache: an assessment of the evidence on clinical diagnosis, invasive tests, and treatment. Lancet Neurol. 2009 Oct; 8(10):959–68.

Facet joints, comprising symmetrical synovial-lined joints enveloped by a fibrous capsule, connect the articular facets of the vertebrae, and facilitate the flexion, extension, and rotation of the cervical spine [39]. The atlantoaxial joint is distinct from the functional units of the lower cervical spine, as it lacks the posterior articulations characteristic of a true zygapophyseal joint. C2-3 facet joints, which are the uppermost facet joints, are susceptible to degenerative changes associated with the normal aging process, with arthropathy being a common occurrence, as approximately 70% of compressive forces from intervertebral discs eventually transfer to the facet joints [40]. Additionally, these joints are vulnerable to traumatic injuries, particularly those resulting from whiplash incidents, which have been extensively studied [41]. Although not definitively proven, it is theorized that excessive flexion/extension, prompting the release of inflammatory cytokines, contributes to joint pathology and nerve irritation [9].

Small medial branch nerves play a crucial role by transmitting afferent impulses to the brain through their corresponding nerve roots, responsible for the sensory innervation of facet joints [42]. The AA and AO joints are innervated by the ventral rami of the first and second cervical nerves. Meanwhile, the C2/3 facet joint receives innervation from branches of the dorsal ramus of the C3 spinal nerves, including the third occipital nerve—a superficial medial branch of the C3 root responsible for nuchal skin sensation [43]. Further, cervical facet joints below the C2/3 level are innervated by medial branches originating from both upper and lower levels of the joint [39]. For instance, sensory impulses from C3/4 facet joints are conveyed by C3 and C4 medial branches (Fig. 7.3). However, there is variation in the facet joint innervation as additional nerve supply is observed from spinal nerves two or three levels distant or from contralateral spinal nerves [44]. The greater and lesser occipital nerves,

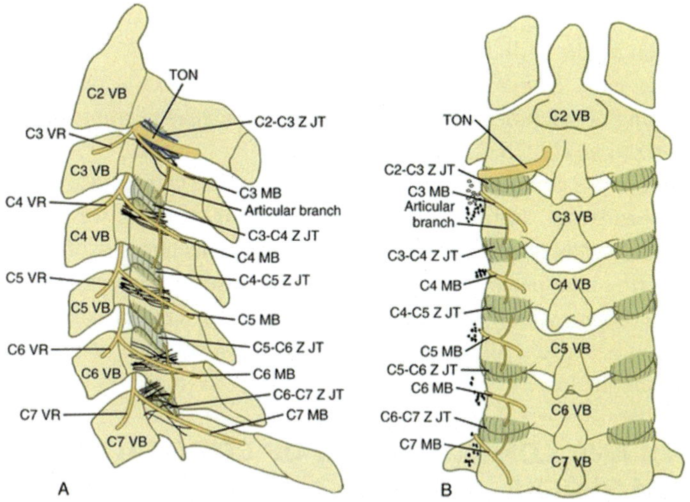

FIGURE 7.3 Location and innervation of cervical medial branches. Derived from: Atlas of Image-Guided Spinal Procedures Book • Second Edition • 2018(FURMAN), Fig 30H.1.

responsible for occipital area sensation, also stem from the upper cervical roots. The greater occipital nerve originates from the ventral ramus of the second cervical nerve, while the lesser occipital arises from the second cervical roots or a combination of the second and third spinal nerves. In cases where peripheral nerve pathology is suspected, occipital nerve blocks can serve diagnostic and therapeutic purposes [6]. Given its origin and shared neural transmission with the cervical medial branch leading to the trigeminal nucleus, the greater occipital nerve block has also shown efficacy in alleviating cervicogenic headaches [45,46].

In diagnosing cervicogenic headaches, employing a confirmatory block guided by imaging is pivotal, as it prevents unwarranted medical expenses incurred by pursuing an inaccurate diagnosis and misguided treatment [17]. If the AA or AO joint is suspected as the pain source, intraarticular injection of local anesthetics (LA) is utilized for joint anesthesia. Below the AA joint, MBBs are employed to obstruct the transmission of sensory information for confirmation [9]. Amide LA is the preferred choice for anaesthetizing medial branches, given its hypoallergenic properties and enhanced tolerability [47]. Amide LA functions by blocking depolarization, thereby interfering with sodium and potassium channels. It exhibits superior infiltration into C fibers compared with A fibers due to differences in myelin thickness. Among amide LA options, bupivacaine, the most potent, offers an extended duration of action, with an onset of action within 8–12 min and lasting 4–8 h. The use of epinephrine is discouraged, and the addition of steroids lacks proven benefits [48].

The suggested mechanism behind acute pain relief involves the reduction of afferent "tone," thereby decreasing the pain generation activity within the trigeminocervical nucleus and cervical dorsal horns. Building upon the convergence phenomenon we previously discussed, reducing pain perception in the cervical area ultimately leads to a broader reduction of pain, extending to areas governed by the trigeminal nucleus [49]. Long-term pain relief following MBBs can be attributed to "breaking the cycle," a concept that involves modulating central sensitization at the initial nociceptive synapse, thereby mitigating persistent neuropathic pain. Moreover, the temporary neural blockade on distinct sets of converging nociceptors has been proposed as a means of improving peripheral sensitization [49,50].

In summary, a comprehensive understanding of the pathophysiology of cervicogenic headaches originating from facet joints is essential for grasping the mechanisms at play during MBB therapy. These blocks not only offer acute relief by altering pain perception but also hold promise for disrupting the persistence of chronic pain through the modulation of central and peripheral sensitization processes, all rooted in the intricate convergence within the trigeminocervical nucleus.

SAFETY AND COMPLICATIONS OF CERVICAL MEDIAL BRANCH BLOCKS

Currently, the Spine Intervention Society (SIS) accepts the dual diagnostic MBB criteria requiring a minimum of $\geq$80% pain relief before progressing to cervical medial branch RFA treatment [17,51,52]. However, these patients inevitably receive RFA and current literature has found no difference in pain outcomes between patients receiving one MBB compared with two, though the time to definitive treatment doubled in patients who received both [53].

Cervical MBBs, conducted according to the SIS's procedure guidelines, is a safe spinal interventional procedure as there are no vulnerable structures in the target region or along the intended track of the needle [51]. Intramedullary injection and ischemic adverse events may occur if the needle strays from its target region [54,55].

A case report documented an American Spinal Injury Association (ASIA)/Frankel grade C injury following an injection of dexamethasone, with allodynia and burning throughout the right C5 dermatome, indicative of an intramedullary injection [54]. Computed tomography (CT) and MRI of the cervical spine revealed the presence of air along interspinous and interlaminar spaces at C4–C5 and a spinal cord contusion within the right hemicord at C4, respectively [54].

Ischemic events may be caused by three hypothesized pathways. Firstly, particulate steroids and nonparticulate steroids combined with local anesthetics may cause crystal embolisms when inadvertently injected into the vertebral artery leading. to upper cervical spinal cord and brainstem infarctions [55,56]. Secondly, direct trauma from needle puncture to the vertebral artery may result in vasospasm or arterial dissection [55]. Thirdly, inadvertent intraarterial injection of air could lead to air emboli resulting in distal ischemia [55]. Generally, intravascular events rarely occur with an incidence rate of 3.9% in all cases applied to the cervical spine [57].

Cervical MBBs also carry the risk of local anesthetic toxicity as lidocaine or bupivacaine are often used diagnostically and therapeutically in conjunction with corticosteroids. Toxicity presents typically as systemic central nervous system or cardiovascular toxicity [55]. Severe complications include transient events such as respiratory depression, hypotension, and Horner's syndrome, and lasting complications including seizures and extrapyramidal symptoms [55]. The risk of toxicity involves a dose-dependent threshold, though reported thresholds have varied greatly among isolated case reports [55].

COMPARISON OF TECHNIQUES OF CERVICAL MBB FOR PAIN SYMPTOMS

Image-guided MBBs with local anesthetic have been the most reliable diagnostic method of identifying facet joint pain. Even so, there's technique variation, specifically in type of local anesthetic and volume of injectate used, and imaging method of choice. Since MBBs are typically used diagnostically to identify patients with cervical facetogenic pain who may respond to subsequent RFA, the quality of MBBs directly impacts the efficacy of this algorithm.

Injectate volumes

For MBBs, the current injectate volume ranges from 0.25 to 2.0 mL, with the most clinically used and recommended injectate volumes being 0.25 and

0.50 mL [58,59]. Studies have shown that high injection volumes (greater than 0.25 mL) leads to a higher rate of inadvertent spread to surrounding untargeted tissue [58,59]. This higher rate of contiguous spread may contribute to a high false positive rate, or low specificity, as the extravasation of the injectate may lead to nociceptive blockade in nearby structures that do not directly contribute to facet innervation [58,59]. Furthermore, studies suggest that accuracy in targeting the correct nerve is not mitigated with higher injectate volumes, but determined more so by needle position, plane, orientation, speed of injection, and anatomical variations [59,60]. Therefore, using a reduced volume of 0.25 mL may suffice for diagnostic purposes. However, differences in injectate volume does not indicate significant differences in pain outcome.

Local anesthetic of choice

Cervical MBBs typically involve an injection cocktail of a local anesthetic and corticosteroid, with the anesthetic of choice being either lidocaine or bupivacaine [61]. Lidocaine relief lasts 1–2 h, while bupivacaine relief lasts 3–8 h [61]. However, studies have shown that lidocaine yields better performance in pain alleviation not only in the early phases due to its faster onset of action, but also in the later phases postintervention whereas bupivacaine only provided pain reduction that lasted half the duration that lidocaine did [61]. Lidocaine is also less expensive and less associated with cardiotoxicity and therefore can be considered to be the anesthetic of choice [61].

Method of imaging

As stated in previous sections, many procedural-associated risks and complications arise when needles stray from their intended path and target. Therefore, close imaging is necessary to accurately and safely locate the proper target, which generally includes C3–C7 medial branches [62]. Traditionally, MBBs were performed under fluoroscopy-guided imaging, specifically with anterior-posterior (AP) and lateral views focusing on localizing C5–C7 [62,63]. The modified swimmer's view, in which one arm is abducted overhead and the other is drawn caudally toward patient's feet, has also recently been introduced as an alternative method for fluoroscopy-guided visualization to account for shoulder obscuration seen in AP and lateral views [63]. CT imaging has been increasingly used as an alternative due to its lower radiation dose and ability to provide precise needle positioning with better visualization of both bone target and surrounding soft tissue [64]. Axial CT views of the vertebra allows for optimal visualization of C3–C7 target points [64]. Typically, CT-guided MBBs are performed while patient is under conscious sedation, prone when targeting C3–C6 and supine when targeting C7/8 levels, and arm positioning is according to patient comfort. The needle is incrementally advanced once target is visualized, and needle position is confirmed with spot images after each incremental adjustment until target is reached [64].

Digital subtraction angiography (DSA) has also been used during cervical MBBs, though because it visualizes vasculature, it is mainly used to mitigate MBB-associated risks such as intravascular injections [65]. DSA functions by subtracting the precontrast image from the postcontrast image and has been shown to detect intravascular injections more accurately and static images [65].

EMERGING ULTRASOUND USAGE IN CERVICAL MEDIAL BRANCH BLOCKS

Cervical MBBs have also recently begun to be performed under ultrasound (US) guidance as it requires fewer equipment and avoids radiation while offering real-time visualization of needle advancement [66].

Technique

Current US-guided cervical MBBs are performed with the biplanar technique, in which the patient is typically supine, head and ipsilateral shoulder rests on a pillow/towel roll, with head turned to the opposite side [67]. The C6 vertebral level may be visualized by placing a linear transducer in the axial plane at the posterior border of the sternocleidomastoid (SCM) muscle at the cricoid cartilage level [67]. Moving rostrally to the bifurcation of the carotid artery marks the C3–C4 level [67]. Needle placement is ascertained in this transverse/axial view and confirmed in the coronal view before needle advancement and injection [67–69]. The short-axis/transverse view allows for better visualization of critical blood vessels that course toward the targeted neuroforamen while the long-axis/coronal view allows for better identification of nerves and visualization of multiple cervical levels [70]. This biplanar confirmation approach thus reduces the risk of collateral damage, especially when targeting C7 [66,71].

Ultrasound versus fluoroscopy

As mentioned, fluoroscopy has been the traditional imaging modality when performing cervical MBBs to the C3–C6 levels. However, fluoroscopy guidance is limited in its ability to clearly visualize C7 structures. Although C7 MBBs can still be achieved under fluoroscopy, studies have shown that under US guidance, fewer needle passes and a shorter performance time (by 157 s) can be achieved [69,72]. C3–C6 blocks performed under the biplanar US technique have also been seen to provide success rates of 97.5%–100% [68,73,74]. Finlayson et al. reported an incidence rate of 7.9% of US detection of blood vessels directly overlaying the targeted articular pillar, which is similar to the rate of unintentional intravascular injection (7%) when performing cervical MBBs under fluoroscopy [73]. However, though detection incidence is not significantly lower when under US guidance, incidence of collateral intravascular damage is lower because under US guidance, Finlayson et al. reported that the operator was able to position the needle without vascular breach and avoided intravascular injection in all cases where overlaying vessels were identified [73].

Compared with fluoroscopy-guided cervical MBBs, pain outcomes from an US-guided cervical MBB are comparable. Park et al. found that success rates at 6-month after the last injection were 55.9% and 51.7% in patients under US-guidance and fluoroscopy-guidance, respectively [72].

Limitations

The major limitation to consistent successful and uncomplicated MBBs under US guidance is skill of the operator, especially in terms of anatomical and needle visualization [72,75]. Improving US proficiency can be achieved through practicing scanning and sonoanatomy identification on oneself and colleagues, practicing needle insertion using simulators, cadavers, or phantoms, and conducting supervised needle placement on patients [75].

CLINICAL STUDIES: SAFETY AND EFFICACY

Currently there is no universal official approach to performing cervical MBBs to maximize safety and decrease adverse effects. With that said, two different approaches are typically done based on the target: a lateral technique for the C3–C7 medial branch nerves and third occipital nerve (TON) and a posterior technique to the C8 medical branch nerve given that it is located at the superolateral aspect of the T1 transverse process [76,77]. Two groups contrast when it comes to the general lateral versus posterior approach. The group that supports the lateral approach reasons that this direction enables the least amount of penetration from skin entry to target site contact thereby decreasing procedure time and risk of tissue damage and ultimately improving patient satisfaction. The other group mentions that the posterior approach is better because it aligns with the goal of radiofrequency (RF) electrode direction, which is what most physicians use and are most comfortable with thereby increasing safety and efficacy.

Regardless, both these approaches are seen as the standard of care based on foundation literature to establish safety and efficacy. Both the safety and efficacy of TON and cervical MBBs must be evaluated. Details of the studies examining the safety and efficacy of cervical MBB and subsequently RFA for cervicogenic headache are listed in Table 7.2.

Safety

Overall, large cohort studies where the clinical standard of care is used (lateral technique for C3-7 MBBs and TON and posterior approach for C8 MBB) show that cervical MBBs are extremely safe [76]. One prospective crossover comparison study showed that the lateral approach clinical standard indicates a practical use because it is faster to perform and enables easier access using a single needle [77]. It also has fewer adverse events such as procedure pain from less overall penetration and thus better post-MBB pain assessments of efficacy, and rarely less need for sedation. One prospective study showed the safety of a TON lateral approach [80]. Similar research evaluating the safety

Table 7.2 Safety of cervical medial branch blocks for cervicogenic headaches.

[76]	Quality assurance databases at three academic interventional pain management practices that utilize evidence-based guidelines were interrogated for immediate complications from interventional pain procedures. 3 academic interventional spine practices Mayo clinic: 61yo, 46% M, 19,170 procedures Rehabilitation Institute of Chicago: 54.5yo, 41.9% M, 6280 procedures University of Pennsylvania: 52.8yo, 47.3% M, 701 procedures	Immediate complication data were available for 26,061 consecutive procedures. A radiology practice performed 19,170 epidural steroid (primarily transforaminal), facet, sacroiliac, and trigger point injections (2006–2013). A physiatry practice performed 6190 spine interventions (2004–09). A second physiatry practice performed 701 spine procedures (2009–2010). There were no major complications (permanent neurologic deficit or clinically significant bleeding [e.g., epidural hematoma]) with any procedure. Overall complication rate was 1.9% (493/26,061). Vasovagal reactions were the most frequent event (1.1%). Nineteen patients (<0.1%) were transferred to emergency departments for: allergic reactions, chest pain, symptomatic hypertension, and a vasovagal reaction.
[77]	In a multicenter setting, 24 subjects underwent 2 separate diagnostic medial-branch blocks in a randomized, single-blind crossover comparison of the single-needle and multiple-needle techniques. Multiple variables were compared between the 2 techniques, including procedure-related discomfort, post-procedure pain relief, volume of local anesthetic required, accuracy as determined by final needle position and contrast-media spread, and time needed to perform the procedure.	The single-needle technique resulted in less procedure-related pain ($P = 0.0003$), required less superficial local anesthesia ($P = 0.0006$), and took less time to complete ($P < 0.0001$) than did the multiple-needle approach. Regarding final needle position, contrast spread, and post-procedure pain relief ($P = 0.8$), no differences were noted between the 2 techniques.
[78]	The records of 613 patients were reviewed who had neck and back pain and were treated with spinal pain blocks between December 2009 and January 2011. The records of patients were reviewed who were treated with spinal pain blocks between December 2009 and January 2011. The types of blocks performed were medial branch blocks, interlaminar epidural blocks and transforaminal epidural blocks. During the first 8 months of the study period (Group A), 2% mepivacaine HCL and triamcinolone was used, and during the last 6 months of the	3/197 patients who received cervical MBB had complications (1.5%). Study did not primarily examine the cervical MBB complication rate but data was given in the study to determine this. Patient 1: 34yo F had a cervical sprain, received C3-7 right MBB and had quadriparesis, chest discomfort, and nausea with a probable cause of underlying conversion disorder. Recovery in 2 months Patient 2: 52yo F had a herniated nucleus pulposus C4-5 and received a C4-6 left MBB. Side effects included respiratory failure, chest discomfort, quadriparesis.

Continued

Table 7.2 Safety of cervical medial branch blocks for cervicogenic headaches. continued

	study period (Group B), mepivacaine was diluted to 1% with normal saline.	Probable cause of inadvertent vascular injection. Recover in 1 h. Patient 3: 59yo F with HNP C5-7 received a C5-7 right MBB. Side effects included chest pain and respiratory depression. The probable cause was inadvertent vascular injection. Recovery in 10 minutes.
[65]	The purpose of this study was to compare the rate of detection of intravascular injections during CMBBs using DSA and static images obtained through conventional fluoroscopy. Methods: 72 patients were included, and a total of 178 CMBBs were performed. The respective incidences of intravascular injections during CMBBs using DSA and static images from conventional fluoroscopy were measured.	A total of 178 CMBBs were performed on 72 patients. All cases of intravascular injections evidenced by the static images were detected by the DSAs. The detection rate of intravascular injections was higher from DSA images than from static images (10.7% vs. 1.7%, $P < .001$).
[59]	Twenty-four patients received cervical medial branch blocks, using either 0.5 or 0.25 mL of bupivacaine mixed with contrast. One-half of the patients in each group were suballocated to receive the blocks in the prone position and the other half through a lateral approach. Participants then underwent computed tomography of the cervical spine to evaluate accuracy and patterns of aberrant contrast spread.	Among the 86 nerve blocks performed, intravascular contrast uptake was noted in six cases, all of which were remedied On the first attempt by slightly repositioning the needle Sixteen instances of aberrant spread were observed in nine patients receiving blocks using 0.5 mL versus seven occurrences in six patients in the 0.25 mL group (P 0.07). The aberrant spread was most observed (57%) when an injection at C3 engulfed the third occipital nerve. Among the 86 nerve blocks, foraminal spread occurred in five instances using 0.5 mL and in two cases with 0.25 mL. The six "missed" nerves were equally divided between treatment groups. No significant difference in any outcome measure was observed between the prone and lateral positions
[79]	Clinical observational study Medial branch blocks were then performed as diagnostic procedures to confirm the zygapophysial joint(s) as the suspected source of pain. Blocks were performed by experienced practitioners on nonidentified patients over a 3-year period. Clinical observations were recorded for 14,312 separate medial branch block levels. The level of the spine and the incidence of inadvertent intravascular injections were recorded.	4134 cervical MBBs that primarily used a lateral approach and occasionally used a posterior approach for the lower cervical levels; they noted a cervical vascular injection rate of 3.9%

Table 7.2 Safety of cervical medial branch blocks for cervicogenic headaches. continued

[29]	Retrospective observational study: 40 patients with refractory cervicogenic headaches and or occipital neuralgia. Patients referred by a headache specialty clinic for evaluation for radiofrequency ablation of the C2 dorsal root ganglion and/or third occipital nerves. Patients were followed for a minimum of 6 months to a year. Patient demographics and the results of radiofrequency ablation were recorded on the same day, after 3 —4 days, and at 6 months to 1 year following treatment.	RFA has a high complication rate between 12% and 13%
[80]	Retrospective chart review Retrospectively analyzed the data of patients who underwent CT-guided RFA of cervical intervertebral discs for cervicogenic headache at the pain medicine Center of Zhejiang Provincial People's Hospital from January 2017 to April 2021. If the conservative treatment failed in patients with discogenic cervicogenic headache, classified the patients as having refractory cervicogenic headache and performed RFA of cervical intervertebral discs	No serious treatment-related complications occurred in this study

of the posterior TON MBB approach to align with the RFA direction has not been published to the best of our knowledge. A couple of studies that used alternate approaches have shown to have a myriad of adverse effects that range from mild to severe; however it is unclear if there is an association between the alternate methods and the adverse effects given this was not described in the studies [78,81]. Further, there is no research published to the best of our knowledge that mentions the lateral and posterior techniques decrease the risk of intravascular adverse injection events such as unintended penetration of the cervical radiculomedullary and vertebral arteries [57,59]. The rate of unintended intravascular injections varies from 2% to 10% when DSA is used but the sample sizes were small in the literature; thus, the confidence in this incidence rate is minimal [57,59,65]. One clinical observational study evaluated 4134 cervical MBBs that primarily used a lateral approach and occasionally used a posterior approach for the lower cervical levels; they noted a cervical vascular injection rate of 3.9% [57]. Lastly, as mentioned before the lateral approach has less overall penetration; but the posterior approach to C8 MBB has more overall penetration. This could prove somewhat troublesome in patients with thicker necks such as obese individuals. This may lead to complications and additional sedation needs.

In the studies that examined RFA adverse effects for cervicogenic headaches, the incidence of the adverse effects was not common. One study found that RFA has a high complication rate between 12% and 13% published in 2014 [29]. However, another study showed a significantly lower adverse event rate, which may indicate the role of advancement and innovation in pain technology as it pertains to RFA efficacy [80].

Efficacy

The cervical facet joints are innervated by the medial branch of the posterior rami that innervates the joints superiorly and inferiorly. Medial Branch RFA (MBRFA) is typically seen as a treatment for patients with neck pain in the context of cervicogenic headaches or other headaches as well. There is some contrasting data about the therapeutic benefits of MBB with LA and/or steroids.

Typically, cervical MBB are diagnostic, and the therapeutic effects last for a transient amount of hours depending on if a short-acting (2 h) or long-acting LA (3–8 h) is used [82–84]. However, sometimes there is an even longer therapeutic benefit, which then makes the typical subsequent RFA treatment not needed. In one prospective cohort study, the researchers examined two groups of shoulder/neck pain and neck pain/headache, where one group of 17 patients had neck pain and headache originating from the C2–C3 joint. They injected LA (bupivacaine 0.5%) as part of the cervical MBB and ultimately 17 of 24 patients had pain relief for a minimum of 2 h and one patient reported headache relief for over 1 month [85]. One study examined MBB and RFA (experimental group) versus MBB and sham RFA (control group) for cervicogenic headache. Specifically, four out of six in the experimental group experienced good outcomes versus two out of six in the control group experienced good outcomes at 3 months posttreatment [86].

Assuming the typical diagnostic use of cervical MBB, clinicians will often follow up with RFA to provide long-lasting relief. In these six studies examining RFA treatment for cervicogenic headaches, five out of the six studies reported pain relief in patient populations greater than or equal to 50% of the patients [24,26,29,30,80,87].

General

Cervical MBB and RFA procedures for cervicogenic headaches have increasingly more relevance as there is a shift toward minimally invasive procedures in the field of interventional pain medicine. In fact, the intersection of RFA and its uses on the nervous system began to create a minimally invasive way to treat various pain pathologies. Further, MBB as a diagnostic tool can sometimes be therapeutic long term, making the use of RFA not needed. However, for most people who undergo cervical MBB and see success will ultimately receive RFA as the MBB is primarily used as a diagnostic tool. While RFA may have a somewhat increased complication rate, most patients who received RFA will elect to repeat the procedure due to its efficacy [84–87].

CONCLUSION

Cervicogenic headaches are a very common cause of headache pain. They are usually unilateral but can sometimes be bilateral in worse cases. Specifically, can originate from the spinal cord or tissue along the neck and trapezius area. It is typically episodic and can radiate throughout the various head areas including the frontal, ocular, occipital, and temporal areas. This can also sometimes be shaped in a C-shaped manner. They can be in a neck-shaped configuration as well. There are various treatment strategies for treating these types of headaches that follow a least invasive to most invasive approach ranging from physiotherapy, anesthetics, and surgical approaches. One of these treatment approaches includes MBBs and if successful, is typically subsequently followed by RFA. They can also be used in conjunction with epidural, steroid injections, and oral anesthetics if needed for a synergistic effect. Efficacy parameters can include, but are not limited to intensity, duration, and complications. Specifically, for RFA, there is a lower complication rate as seen in the literature, specifically when using images of guided procedures, including fluoroscopy or CT. With advancing technology that is constantly evolving, simulations and emerging technologies will enable the practitioner to have increased accuracy and subsequently improved patient outcomes when using RFA.

REFERENCES

[1] E. Anarte-Lazo, G.F. Carvalho, A. Schwarz, K. Luedtke, D. Falla, Differentiating migraine, cervicogenic headache and asymptomatic individuals based on physical examination findings: a systematic review and meta-analysis, BioMed Central Ltd, United Kingdom BMC Musculoskeletal Disorders 22 (1) (2021), https://doi.org/10.1186/s12891-021-04595-w. Available from: http://www.biomedcentral.com/bmcmusculoskeletdisord/.

[2] S. Haldeman, S. Dagenais, Cervicogenic headaches, Spine J. 1 (1) (2001) 31–46, https://doi.org/10.1016/s1529-9430(01)00024-9.

[3] F. Antonaci, G. Bono, P. Chimento, Diagnosing cervicogenic headache, J. Headache Pain 7 (3) (2006) 145–148, https://doi.org/10.1007/s10194-006-0277-3.

[4] J.N. Blau, E.A. MacGregor, Migraine and the neck, Headache J. Head Face Pain 34 (2) (1994) 88–90, https://doi.org/10.1111/j.1526-4610.1994.hed3402088.x.

[5] T.A. Fredriksen, F. Antonaci, O. Sjaastad, Cervicogenic headache: too important to be left un-diagnosed, Springer-Verlag Italia s.r.l., Norway, J. Headache Pain 16 (1) (2015), https://doi.org/10.1186/1129-2377-16-6. Available from: https://thejournalofheadacheandpain.biomedcentral.com/.

[6] Headache classification committee of the international headache society (IHS) the international classification of headache disorders, 3rd edition, Cephalalgia 38 (1) (2018) 1–211, https://doi.org/10.1177/0333102417738202.

[7] O. Sjaastad, T.A. Fredriksen, V. Pfaffenrath, Cervicogenic headache: diagnostic criteria, Headache J. Head Face Pain 38 (6) (1998) 442–445, https://doi.org/10.1046/j.1526-4610.1998.3806442.x.

[8] N. Bogduk, Cervicogenic headache: anatomic basis and pathophysiologic mechanisms, Curr. Pain Headache Rep. 5 (4) (2001) 382–386, https://doi.org/10.1007/s11916-001-0029-7.

[9] J.D. Hoppenfeld, Cervical facet arthropathy and occipital neuralgia: headache culprits, Curr. Pain Headache Rep. 14 (6) (2010) 418–423, https://doi.org/10.1007/s11916-010-0151-5.

[10] G. Cooper, B. Bailey, N. Bogduk, Cervical zygapophysial joint pain maps, Pain Med. 8 (4) (2007) 344–353, https://doi.org/10.1111/j.1526-4637.2006.00201.x.

[11] O. Coskun, S. Ucler, B. Karakurum, H.T. Atasoy, T. Yildirim, S. Ozkan, L.E. Inan, Magnetic resonance imaging of patients with cervicogenic headache, Cephalalgia 23 (8) (2003) 842–845, https://doi.org/10.1046/j.1468-2982.2003.00605.x.

[12] J. Govind, N. Bogduk, Sources of cervicogenic headache among the upper cervical synovial joints, Pain Med. 23 (6) (2022) 1059–1065, https://doi.org/10.1093/pm/pnaa469.

[13] N. Bogduk, C. Aprill, On the nature of neck pain, discography and cervical zygapophysial joint blocks, Pain 54 (2) (1993) 213–217, https://doi.org/10.1016/0304-3959(93)90211-7.

[14] S.W. Park, Y.S. Park, T.K. Nam, T.G. Cho, The effect of radiofrequency neurotomy of lower cervical medial branches on cervicogenic headache, Korean Neurosurgical Society, South Korea, Journal of Korean Neurosurgical Society 50 (6) (2011) 507–511, https://doi.org/10.3340/jkns.2011.50.6.507. Available from: http://www.jkns.or.kr/htm/pdfdown.asp?pn=0042011182.

[15] R. Michler, G. Bovim, O. Sjaastad, Disorders in the lower cervical spine. A cause of unilateral headache? A case report, Headache J. Head Face Pain 31 (8) (1991) 550–551, https://doi.org/10.1111/j.1526-4610.1991.hed3108550.x.

[16] J.B. Lee, J.Y. Park, J. Park, D.J. Lim, S.D. Kim, H.S. Chung, Clinical efficacy of radiofrequency cervical zygapophyseal neurotomy in patients with chronic cervicogenic headache, J. Kor. Med. Sci. 22 (2) (2007) 326, https://doi.org/10.3346/jkms.2007.22.2.326.

[17] R.W. Hurley, M.C.B. Adams, M. Barad, A. Bhaskar, A. Bhatia, A. Chadwick, T.R. Deer, J. Hah, W.M. Hooten, N.R. Kissoon, D.W. Lee, Z. McCormick, J.Y. Moon, S. Narouze, D.A. Provenzano, B.J. Schneider, M. van Eerd, J. Van Zundert, M.S. Wallace, S.M. Wilson, Z. Zhao, S.P. Cohen, Consensus practice guidelines on interventions for cervical spine (facet) joint pain from a multispecialty international working group, BMJ Publishing Group, United States, Reg. Anesth. Pain Med. 47 (1) (2022) 3–59, https://doi.org/10.1136/rapm-2021-103031.

[18] J.P. Paquin, Y. Tousignant-Laflamme, J.P. Dumas, Effects of SNAG mobilization combined with a self-SNAG home-exercise for the treatment of cervicogenic headache: a pilot study, Taylor and Francis Ltd., Canada, J. Man. Manip. Ther. 29 (4) (2021) 244–254, https://doi.org/10.1080/10669817.2020.1864960.

[19] N. Bogduk, J. Govind, Cervicogenic headache: an assessment of the evidence on clinical diagnosis, invasive tests, and treatment, Lancet Neurol. 8 (10) (2009) 959–968, https://doi.org/10.1016/s1474-4422(09)70209-1.

[20] N. Nilsson, A randomized controlled trial of the effect of spinal manipulation in the treatment of cervicogenic headache, J. Manipulative Physiol. Therapeut. 18 (7) (1995) 435–440.

[21] N. Nilsson, H.W. Christensen, J. Hartvigsen, The effect of spinal manipulation in the treatment of cervicogenic headache, J. Manipulative Physiol. Therapeut. 20 (5) (1997) 326–330.

[22] H.T. Vernon, Spinal manipulation and headaches of cervical origin, J. Manipulative Physiol. Therapeut. 12 (6) (1989) 455–468.

[23] M. Fernandez, C. Moore, J. Tan, D. Lian, J. Nguyen, A. Bacon, B. Christie, I. Shen, T. Waldie, D. Simonet, A. Bussières, Spinal manipulation for the management of cervicogenic headache: a systematic review and meta-analysis, Eur. J. Pain 24 (9) (2020) 1687–1702, https://doi.org/10.1002/ejp.1632.

[24] S.R.S. Haspeslagh, H.A. Van Suijlekom, I.E. Lamé, A.G.H. Kessels, M.van Kleef, W.E.J. Weber, Randomised controlled trial of cervical radiofrequency lesions as a

treatment for cervicogenic headache [ISRCTN07444684], BMC Anesthesiol. 6 (1) (2006), https://doi.org/10.1186/1471-2253-6-1.
[25] J. Govind, Radiofrequency neurotomy for the treatment of third occipital headache, J. Neurol. Neurosurg. Psychiatr. 74 (1) (2003) 88–93, https://doi.org/10.1136/jnnp.74.1.88.
[26] Lee, H.H. Cho, F.S. Nahm, B. Lee, E. Choi, Pulsed radiofrequency ablation of the C2 dorsal root ganglion using a posterior approach for treating cervicogenic headache: a retrospective chart review, Headache J. Head Face Pain 60 (10) (2020) 2463–2472, https://doi.org/10.1111/head.13759.
[27] S.j. Li, D. Feng, Pulsed radiofrequency of the C2 dorsal root ganglion and epidural steroid injections for cervicogenic headache, Springer-Verlag Italia s.r.l., China Neurological Sciences 40 (6) (2019) 1173–1181, https://doi.org/10.1007/s10072-019-03782-x.
[28] J.F. Hamer, T.A. Purath, Repeat RF ablation of C2 and third occipital nerves for recurrent occipital neuralgia and cervicogenic headaches, World J. Neurosci. 06 (04) (2016) 236–242, https://doi.org/10.4236/wjns.2016.64029.
[29] J.F. Hamer, T.A. Purath, Response of cervicogenic headaches and occipital neuralgia to radiofrequency ablation of the C2 dorsal root ganglion and/or third occipital nerve, Headache J. Head Face Pain 54 (3) (2014) 500–510, https://doi.org/10.1111/head.12295.
[30] W. Halim, N.H.L. Chua, K.C. Vissers, Long-term pain relief in patients with cervicogenic headaches after pulsed radiofrequency application into the lateral atlantoaxial (C1-2) joint using an anterolateral approach, Pain Pract. 10 (4) (2010) 267–271, https://doi.org/10.1111/j.1533-2500.2010.00360.x.
[31] T. Bartsch, Increased responses in trigeminocervical nociceptive neurons to cervical input after stimulation of the dura mater, Brain 126 (8) (2003) 1801–1813, https://doi.org/10.1093/brain/awg190.
[32] T. Bartsch, Stimulation of the greater occipital nerve induces increased central excitability of dural afferent input, Brain 125 (7) (2002) 1496–1509, https://doi.org/10.1093/brain/awf166.
[33] Bogduk, T. Bartsch, Cervicogenic Headache Wolff 's Headache, eighth ed., Oxford University Press, Oxford University Press, 2008, p. 2008.
[34] N. Bogduk, The neck and headaches, Neurol. Clin. 22 (1) (2004) 151–171, https://doi.org/10.1016/s0733-8619(03)00100-2.
[35] P.J. Goadsby, T. Bartsch, Introduction: on the functional neuroanatomy of neck pain, Cephalalgia 28 (1_Suppl. l) (2008) 1–7, https://doi.org/10.1111/j.1468-2982.2008.01606.x.
[36] D.G. Campbell, C.M. Parsons, Referred head pain and its concomitants: report of preliminary experimental investigation with implications for the post-traumatic "head" syndrome, J. Nerv. Ment. Dis. 99 (5) (1944) 544–551, https://doi.org/10.1097/00005053-194405000-00009.
[37] A. Dwyer Mb, C. Aprill, N. Bogduk, Cervical zygapophyseal joint pain patterns I, Spine 15 (6) (1990) 453–457, https://doi.org/10.1097/00007632-199006000-00004.
[38] P. Dreyfuss, M. Michaelsen, D. Fletcher, Atlanto-occipital and lateral atlanto-axial joint pain patterns, Spine 19 (Suppl. ment) (1994) 1125–1131, https://doi.org/10.1097/00007632-199405001-00005.
[39] N. Bogduk, S.M. Lord, Cervical zygapophysial joint pain, Neurosurg. Q. 8 (2) (1998) 107–117, https://doi.org/10.1097/00013414-199806000-00004.
[40] M.C. Lynch, J.F. Taylor, Facet joint injection for low back pain. A clinical study, Journal of Bone and Joint Surgery - Series B 68 (1) (1986) 138–141, https://doi.org/10.1302/0301-620x.68b1.2934398.
[41] S.M. Lord, L. Barnsley, B.J. Wallis, N. Bogduk, Chronic cervical zygapophysial joint pain after whiplash: a placebo- controlled prevalence study, Spine 21 (15) (1996) 1737–1745, https://doi.org/10.1097/00007632-199608010-00005.
[42] T. Kweon, J. Kim, H. Lee, M. Kim, Y.W. Lee, Anatomical analysis of medial branches of dorsal rami of cervical nerves for radiofrequency thermocoagulation, Lippincott

Williams and Wilkins, South Korea, Reg. Anesth. Pain Med. 39 (6) (2014) 465–471, https://doi.org/10.1097/AAP.0000000000000175.
[43] N. Bogduk, A. Marsland, On the concept of third occipital headache, J. Neurol. Neurosurg. Psychiatr. 49 (7) (1986) 775–780, https://doi.org/10.1136/jnnp.49.7.775.
[44] N. Bogduk, J. Macintosh, A. Marsland, Technical limitations to the efficacy of radiofrequency neurotomy for spinal pain, Neurosurgery 20 (4) (1987) 529–535, https://doi.org/10.1227/00006123-198704000-00004.
[45] G. Bovim, Cervicogenic headache, migraine, and tension-type headache. Pressure-pain threshold measurements, Pain 51 (2) (1992) 169–173, https://doi.org/10.1016/0304-3959(92)90258-d.
[46] P. Rothbart, K. Fiedler, G.D. Gale, D. Nussbaum, N. Hendlerb, A descriptive study of 100 patients undergoing palliative nerve blocks for chronic intractable headache and neck ache, Pain Res. Manag. 5 (4) (2000) 243–248, https://doi.org/10.1155/2000/706434.
[47] M. Levin, Nerve blocks in the treatment of headache, Neurotherapeutics 7 (2) (2010) 197–203, https://doi.org/10.1016/j.nurt.2010.03.001.
[48] A. Ashkenazi, R. Matro, J.W. Shaw, M.A. Abbas, S.D. Silberstein, Greater occipital nerve block using local anaesthetics alone or with triamcinolone for transformed migraine: a randomised comparative study, J. Neurol. Neurosurgery Psychiatry 79 (4) (2008) 415–417, https://doi.org/10.1136/jnnp.2007.124420.
[49] E.J. Piovesan, P.A. Kowacs, C.E. Tatsui, M.C. Lange, L.C. Ribas, L.C. Werneck, Referred pain after painful stimulation of the greater occipital nerve in humans: evidence of convergence of cervical afferences on trigeminal nuclei, Cephalalgia 21 (2) (2001) 107–109, https://doi.org/10.1046/j.1468-2982.2001.00166.x.
[50] V. Busch, W. Jakob, T. Juergens, W. Schulte-Mattler, H. Kaube, A. May, Functional connectivity between trigeminal and occipital nerves revealed by occipital nerve blockade and nociceptive blink reflexes, Cephalalgia 26 (1) (2006) 50–55, https://doi.org/10.1111/j.1468-2982.2005.00992.x.
[51] Bogduk N International Spine Intervention Society, Practice Guidelines for Spinal Diagnostic and Treatment Procedures, second ed., International Spine Intervention Society, San Francisco, 2013, p. 2013.
[52] D. Sherwood, E. Berlin, A. Epps, J. Gardner, B.J. Schneider, Cervical medial branch block progression to radiofrequency neurotomy: a retrospective clinical audit, North American Spine Soc J (NASSJ) 8 (2021) 100091, https://doi.org/10.1016/j.xnsj.2021.100091.
[53] J.B. Cadwell, M.A. Gaballa, S. Jain, S. Patel, P. Hesketh, A. Adaralegbe, S. Char, A.G. Kaufman, One block too much? Pain outcomes for patients receiving one versus two medial branch blocks before thermal radiofrequency ablation, Reg. Anesth. Pain Med. (2023), https://doi.org/10.1136/rapm-2023-104457.
[54] A. Miller, D. Griepp, R. Rahme, Beware the wandering needle: inadvertent intramedullary injection during an attempted cervical medial branch block, World Neurosurgery 149 (2021) 169–170, https://doi.org/10.1016/j.wneu.2021.02.107.
[55] Z.E. Stewart, Safety of local anesthetics in cervical nerve root injections: a narrative review, Springer Science and Business Media Deutschland GmbH, United States, Skeletal Radiol. 52 (10) (2023) 1893–1900, https://doi.org/10.1007/s00256-022-04220-4.
[56] E.J. Choi, D.H. Kim, W.K. Han, H.J. Lee, I. Kang, F.S. Nahm, P.B. Lee, Non-particulate steroids (Betamethasone sodium phosphate, dexamethasone sodium phosphate, and dexamethasone palmitate) combined with local anesthetics (ropivacaine, levobupivacaine, bupivacaine, and lidocaine): a potentially unsafe mixture, Dove Medical Press Ltd, South Korea, J. Pain Res. 14 (2021) 1495–1504, https://doi.org/10.2147/JPR.S311573.
[57] P. Verrills, B. Mitchell, D. Vivian, G. Nowesenitz, B. Lovell, C. Sinclair, The incidence of intravascular penetration in medial branch blocks: cervical, thoracic, and lumbar spines, Spine 33 (6) (2008) E174, https://doi.org/10.1097/BRS.0b013e318166f03d.
[58] S.E. Wahezi, J.J. Molina, E. Alexeev, J.S. Georgy, N. Haramati, S.A. Erosa, J.M. Shah, S. Downie, Cervical medial branch block volume dependent dispersion patterns as a

predictor for ablation success: a cadaveric study, John Wiley and Sons Inc., United States, PM and R 11 (6) (2019) 631–639, https://doi.org/10.1016/j.pmrj.2018.10.003.

[59] S.P. Cohen, S.A. Strassels, C. Kurihara, A. Forsythe, C.C. Buckenmaier, B. McLean, G. Riedy, S. Seltzer, Randomized study assessing the accuracy of cervical facet joint nerve (medial branch) blocks using different injectate volumes, Lippincott Williams and Wilkins, United States, Anesthesiology 112 (1) (2010) 144–152, https://doi.org/10.1097/ALN.0b013e3181c38a82.

[60] S. Datta, L. Manchikanti, Different injectate volumes in cervical medial branch blocks: does increased diagnostic accuracy with smaller injectate volume lead to changes in outcome? Anesthesiology 113 (2) (2010) 499, https://doi.org/10.1097/ALN.0b013e3181e6a2b8.

[61] P. Pasuhirunnikorn, T. Tanasansomboon, W. Singhatanadgige, W. Yingsakmongkol, P. Chalermkitpanit, Comparative outcome of lidocaine versus bupivacaine for cervical medial branch block in chronic cervical facet arthropathy: a randomized double-blind study, World Neurosurgery 175 (2023) e662, https://doi.org/10.1016/j.wneu.2023.04.003.

[62] J. Artner, S. Klessinger, Interventionen an facettengelenken, Radiologe 55 (10) (2015) 840–846, https://doi.org/10.1007/s00117-015-0006-5.

[63] D.S. Stolzenberg, R. Pfeifer, J. Armstrong, G. Young, J. Gehret, J. Simon, G.C.C. Chien, Improving fluoroscopic visualization for lower cervical medial branch blocks with a modified swimmer's view: a technical report, American Society of Interventional Pain Physicians, United States, Pain Physician 21 (3) (2018) 303–308, https://doi.org/10.36076/ppj.2018.3.303.

[64] M. Khan, S. Meleka, CT guided cervical medial branch block and radiofrequency ablation, J. Clin. Neurosci. 78 (2020) 393–396, https://doi.org/10.1016/j.jocn.2020.05.009.

[65] Y.H. Jeon, S.Y. Kim, Detection rate of intravascular injections during cervical medial branch blocks: a comparison of digital subtraction angiography and static images from conventional fluoroscopy, Korean Pain Society, South Korea, Korean Journal of Pain 28 (2) (2015) 105–108, https://doi.org/10.3344/kjp.2015.28.2.105.

[66] D. Park, M.Y. Seong, H.Y. Kim, J.S. Ryu, Spinal cord injury during ultrasound-guided C7 cervical medial branch block, Lippincott Williams and Wilkins, South Korea, Am. J. Phys. Med. Rehab. 96 (6) (2017) e111, https://doi.org/10.1097/PHM.0000000000000613.

[67] J. Li, A. Szabova, Ultrasound-guided nerve blocks in the head and neck for chronic pain management: the anatomy, sonoanatomy, and procedure, American Society of Interventional Pain Physicians, United States Pain Physician 24 (8) (2021) 533–548.

[68] R.J. Finlayson, J.P.B. Etheridge, W. Tiyaprasertkul, B. Nelems, D.Q.H. Tran, A prospective validation of biplanar ultrasound imaging for C5-C6 cervical medial branch blocks, Reg. Anesth. Pain Med. 39 (2) (2014) 160–163, https://doi.org/10.1097/AAP.0000000000000043.

[69] R.J. Finlayson, J.P.B. Etheridge, W. Tiyaprasertkul, B. Nelems, D.Q.H. Tran, A randomized comparison between ultrasound- and fluoroscopy-guided C7 medial branch block, Lippincott Williams and Wilkins, Canada, Reg. Anesth. Pain Med. 40 (1) (2015) 52–57, https://doi.org/10.1097/AAP.0000000000000186.

[70] S.N. Narouze, Ultrasound-guided cervical spine injections: ultrasound "prevents" whereas contrast fluoroscopy "detects" intravascular injections, Reg. Anesth. Pain Med. 37 (2) (2012) 127–130, https://doi.org/10.1097/AAP.0b013e31823f3c80.

[71] K.V. Chang, W.T. Wu, L. Özçakar, Ultrasound-guided C7 cervical medial branch block using the in-plane approach, Lippincott Williams and Wilkins, Taiwan, Am. J. Phys. Med. Rehab. 96 (9) (2017) e164, https://doi.org/10.1097/PHM.0000000000000696.

[72] K.D. Park, D.-J. Lim, W.Y. Lee, J.K. Ahn, Y. Park, Ultrasound versus fluoroscopy-guided cervical medial branch block for the treatment of chronic cervical facet joint pain: a retrospective comparative study, Skeletal Radiol. 46 (1) (2017) 81–91, https://doi.org/10.1007/s00256-016-2516-2.

[73] R.J. Finlayson, G. Gupta, M. Alhujairi, S. Dugani, D.Q.H. Tran, Cervical medial branch block: a novel technique using ultrasound guidance, Reg. Anesth. Pain Med. 37 (2) (2012) 219−223, https://doi.org/10.1097/AAP.0b013e3182374e24.

[74] B. Moreno, J. Barbosa, Ultrasound-guided procedures in the cervical spine, Cureus (2021), https://doi.org/10.7759/cureus.20361.

[75] S.Y. Kwon, J.-W. Kim, M.J. Cho, A. Hussain Al-Sinan, Y.-J. Han, Y.H. Kim, The efficacy of cervical spine phantoms for improving resident proficiency in performing ultrasound-guided cervical medial branch block, Medicine 97 (51) (2018) e13765, https://doi.org/10.1097/md.0000000000013765.

[76] C.M. Carr, C.T. Plastaras, M.J. Pingree, M. Smuck, T.P. Maus, J.R. Geske, C.A. El-Yahchouchi, Z.L. McCormick, D.J. Kennedy, Immediate adverse events in interventional pain procedures: a multi-institutional study, Oxford University Press, United States, Pain Med. 17 (12) (2016) 2155−2161, https://doi.org/10.1093/pm/pnw051.

[77] M.P. Stojanovic, D. Dey, E.D. Hord, Y. Zhou, S.P. Cohen, A prospective crossover comparison study of the single-needle and multiple-needle techniques for facet-joint medial branch block, Reg. Anesth. Pain Med. 30 (5) (2005) 484−490, https://doi.org/10.1016/j.rapm.2005.05.007.

[78] H.I. Lee, Y.S. Park, T.G. Cho, S.W. Park, J.T. Kwon, Y.B. Kim, Transient adverse neurologic effects of spinal pain blocks, J. Korean Neurosurg. Soc. 52 (3) (2012) 228−233, https://doi.org/10.3340/jkns.2012.52.3.228.

[79] P. Verrills, B. Mitchell, D. Vivian, G. Nowesenitz, B. Lovell, C. Sinclair, The incidence of intravascular penetration in medial branch blocks: cervical, thoracic, and lumbar spines, Spine (Phila Pa 1976). 33 (6) (2008) E174−7. http://doi.org/10.1097/BRS.0b013e318166f03d. PMID: 18344846.

[80] J.Q. Hu, J. Zhang, B. Ru, W.J. Cai, W.L. Liu, R. Guo, Z.W. Ji, Q. Wan, L.H. Xu, Y. Cheng, J. Zhang, S. Li, Computed tomography-guided radiofrequency ablation of cervical intervertebral discs for the treatment of refractory cervicogenic headache: a retrospective chart review, John Wiley and Sons Inc, China Headache 62 (7) (2022) 839−847, https://doi.org/10.1111/head.14361. Available from: http://onlinelibrary.wiley.com/journal/10.1111/(ISSN)1526-4610.

[81] R.J. Finlayson, J.P.B. Etheridge, L. Vieira, G. Gupta, D.Q.H. Tran, A randomized comparison between ultrasound- and fluoroscopy-guided third occipital nerve block, Reg. Anesth. Pain Med. 38 (3) (2013) 212−217, https://doi.org/10.1097/AAP.0b013e31828b25bc.

[82] M.J. Cousins, L.E. Mather, Clinical pharmacology of local anaesthetics, Anaesth. Intensive Care 8 (3) (1980) 257−277, https://doi.org/10.1177/0310057x8000800303.

[83] S.M. Lord, L. Barnsley, N. Bogduk, The utility of comparative local anesthetic blocks versus placebo- controlled blocks for the diagnosis of cervical zygapophysial joint pain, Lippincott Williams and Wilkins, Australia Clinical Journal of Pain 11 (3) (1995) 208−213, https://doi.org/10.1097/00002508-199509000-00008.

[84] Les Barnsley, S. Lord, N. Bogduk, Comparative local anaesthetic blocks in the diagnosis of cervical zygapophysial joint pain, Pain 55 (1) (1993) 99−106, https://doi.org/10.1016/0304-3959(93)90189-v.

[85] N. Bogduk, A. Marsland, The cervical zygapophysial joints as a source of neck pain, Spine 13 (6) (1988) 610−617, https://doi.org/10.1097/00007632-198806000-00003.

[86] L.J. Stovner, F. Kolstad, G. Helde, Radiofrequency denervation of facet joints C2-C6 in cervicogenic headache: a randomized, double-blind, sham-controlled study, Cephalalgia 24 (10) (2004) 821−830, https://doi.org/10.1111/j.1468-2982.2004.00773.x.

[87] C.A. Odonkor, T. Tang, D. Taftian, A. Chhatre, Bilateral intra-articular radiofrequency ablation for cervicogenic headache, Case Reports in Anesthesiology 2017 (2017) 1−6, https://doi.org/10.1155/2017/1483279.

Chapter | Eight

Cervical radiofrequency ablation—Cervicogenic headaches

Joshua S. Kim[1], Richard W. Kim[2], Aila Malik[3], Peter D. Vu[3]

[1]Weill Cornell Medical College, New York, NY, United States; [2]Columbia University Department of Rehabilitation and Regenerative Medicine, Weill Cornell Department of Rehabilitation Medicine, New York, NY, United States; [3]Department of Physical Medicine & Rehabilitation the University of Texas Health Science Center, McGovern Medical School, Houston, TX, United States

INTRODUCTION

Cervical nerve radiofrequency ablation (RFA) is a frequently utilized procedure performed by interventional pain physicians involving tissue destruction of the afferent nerve supply of the cervical facet joints [1]. This technique is commonly utilized to address cervicogenic headache, which is thought to be mediated by cervical spine pathology [2]. Tissue destruction through RFA is directed at the medial branch of the dorsal primary ramus of the suspected pain generating facet joint(s), thought to be the main source of discomfort for patients suffering from cervicogenic headache [3]. This tissue destruction occurs by a lesion created using radiofrequency energy, causing Wallerian degeneration of the afferent nerve fibers and reducing referred pain [4].

Cervicogenic headache, commonly characterized by exacerbation of headache pain with certain active neck movements, restrictions in cervical range of motion, and association with objective findings of cervical spine pathology, is considered to make up 15%–20% of chronic headache cases and has an overall estimated prevalence of 1%–4.1% in the general population [5,6].For patients suffering from a cervicogenic headache, RFA of the cervical nerves offers a valuable path to pain relief and is a treatment option frequently deployed by pain physicians.

Radiofrequency ablation for cervicogenic pain has a history dating back to 1931 when Kirschner treated Trigeminal Neuralgia using radiofrequency abla-

Interventional Management of Migraines and Other Headache Disorders. https://doi.org/10.1016/B978-0-443-23557-3.00008-5

tion of the Gasserian ganglion [7]. While initially used for Trigeminal Neuralgia, RFA generators became commercially available in the 1950s. [8].Over time, RFA was increasingly investigated and employed to target numerous diverse pain conditions, including thoracic and lumbar facet mediated pain, thoracic radicular pain, and sciatic back pain [9–16]. However, it was not until 1996 that efficacy data regarding use of cervical RFA to treat cervical facet mediated pain was published [17]. The development of alternative RFA methods, such as pulsed and cooled RFA, further broadened the application and the versatility of this treatment modality [18–22].

Current literature shows that cervical facet joint mediated pain, particularly chronic neck pain and neck pain following whiplash injury, holds a high prevalence rate in various regions. A systematic review reported neck pain has a high prevalence rate in both developed and undeveloped regions, particularly in the United States, Western Europe, East Asia, North Africa, and the Middle East, with annual and lifetime neck pain prevalence rates of 37.2% (range 16.7% –75.1%) and 48.5% (range 14.2%–71%), respectively [23]. As previously noted, the prevalence of cervicogenic headache in the general population can range from 1% to 4.1% [6]. This wide disparity in reported prevalence raises questions about the use and accuracy of historical and physical exam signs as diagnostic reference standards. For this reason, it may be difficult to establish a cervical facet joint pathology as the primary source of headache pain in patients with concurrent chronic neck pain and other conditions contributes to this complexity.

Despite debate, cervical nerve RFA continues to evolve as a specialized intervention for chronic intractable headache pain in cases where conservative therapies have failed. This treatment has proven to be highly effective in reducing pain intensity, with many patients experiencing significant benefit, and some achieving complete pain relief after the intervention [24]. The integration of RFA with advanced intraprocedure imaging enhances its impact on reducing pain intensity, as it allows precise targeting and complete ablation of the source of referred pain [2,25]. Patients seeking more than 50% pain relief can reliably achieve this outcome within 6 months to 1 year [1,25,26]. However, the clinical benefit following cervical nerve RFA may be short lived, likely due to nerve regeneration. The average pain improvement duration following an RFA intervention is reported to be approximately 22.5 weeks [25]. Ongoing research and technological innovations suggest likely continual application of this treatment modality in the management of chronic headache.

INDICATIONS FOR CERVICAL NERVE RADIOFREQUENCY ABLATION

A cervicogenic headache is a chronic, recurrent headache that is provoked by movements of the head and neck [27]. It is typically described as being unilateral, nonthrobbing in nature, and may be associated with restricted range of motion of the cervical spine. The headache is caused by a nociceptive source in the cervical spine and pain may be referred to the frontal, temporal, or occipital regions. Referral of pain occurs due to physiological convergence of afferent input from the upper cervical spinal nerves (C1–C3) and the trigeminal nerve within the Trigeminocervical Complex (TCC), a spinal nucleus of the trigeminal nerve extending from the medulla to the upper cervical segments [27,28]. The most caudal portion of the spinal trigeminal nucleus, the pars caudalis, is continuous with the outer lamina of the dorsal horns of the upper cervical segments and receives afferents from the trigeminal nerve and the C1–C3 spinal nerves. This structural overlap of nociceptive input provides the neuroanatomical basis for the frequent cooccurrence of head and neck pain and the bidirectional referral of pain between these regions. Any neck structure innervated by the C1–C3 spinal nerves may be the source of referred pain to the head including the atlantoaxial (AA) joints, cervical facet joints, cervical intervertebral discs, and cervical musculature. The most common source is the C2–C3 facet joint, accounting for up to 70% of cases and resulting in a condition known as the third occipital headache [29]. Although considered a separate entity, occipital neuralgia may be associated with cervicogenic headache, and its presence may worsen cervicogenic headache symptomology [3]. Occipital neuralgia is characterized by paroxysmal lancinating pain in the distribution of the greater or lesser occipital nerves and may result from nerve entrapment [30].

Management of cervicogenic headaches is often multimodal consisting of physical therapy and medical management, which includes use of nonsteroidal antiinflammatory drugs (NSAIDs), muscle relaxants, tricyclic antidepressants, and antiepileptic agents [3]. Interventional treatment options include anesthetic and/or steroid injections to the cervical facet joints, AA joints, and the greater and lesser occipital nerves [31]. Trigger point injections, and botulinum toxin injections may be used in the management of myofascial pain [32]. RFA may be considered in cases where conservative management fails to provide relief. This procedure involves application of radiofrequency currents near nociceptive pathways to modify pain signaling [33]. RFA of the third occipital nerve (TON) supplying the C2–C3 facet joint and medial branch of the cervical dorsal rami is used to treat facet joint mediated pain, a major contributor to cervicogenic headache symptomatology. Other potential targets of this intervention include the C2 dorsal root ganglion, the AA joints, and the greater and lesser occipital nerves [3]. RFA is a minimally invasive procedure and offers an alternative to surgical treatments such as ganglionectomy, surgical transection of the greater occipital nerve, and dorsal decompressive laminectomy.

Etiology

A Cervicogenic headache results from referred pain from cervical structures innervated by the upper three cervical spinal nerves. Possible sources of cervicogenic headache include the atlantooccipital (AO) joint, atlantoaxial (AA) joints, cervical facet joints, cervical intervertebral discs, and the upper cervical spinal nerves and roots. Other serious causes of occipital headaches should be ruled out, such as posterior cranial fossa lesions and vertebral artery dissection or aneurysm [34].

Tumors, fractures, infections, and rheumatoid arthritis of the upper cervical spine have not been validated formally as causes of cervicogenic headache but are nevertheless accepted as valid causes in individual cases. However, cervical spondylosis and osteochondritis are not accepted as valid causes of cervicogenic headache. Also, when myofascial tender points are the cause, the headache should be coded under a type of tension headache [34].

The clinical criteria for cervicogenic headache, as defined by the International Headache Society (IHS), necessitate the presence of at least two indicative factors. Firstly, there should be clear evidence of a temporal relationship between the onset of the headache and the development of a cervical disorder or the appearance of a lesion. Additionally, a notable improvement or resolution of the headache should correspond with the parallel improvement or resolution of the underlying cervical disorder or lesion. Furthermore, there should be a demonstrable reduction in cervical range of motion, and provocative maneuvers, which significantly exacerbate the headache. Finally, the diagnostic blockade of a cervical structure or its nerve supply should result in the complete resolution of the headache, providing additional confirmation of the cervicogenic origin [35].

CLINICAL CRITERIA

The Cervicogenic Headache International Study Group (CHISG) has listed more specific diagnostic criteria [36]. A confirmatory diagnosis is based on a combination of signs and symptoms including neck involvement (precipitation of head pain by neck movement and/or reduced neck range of motion), a positive diagnostic anesthetic block of the suspected cervical structure(s) mediating pain, and unilaterality (although bilateral cases are acceptable). Both the ICH and CHISG diagnostic criteria state the use of local anesthetic blocks of the cervical nerves or cervical facet joints performed at the probable cervical levels as a helpful confirmatory tool to establish a cervical source of headache. Invasive anesthetic blocks, especially when performed at the atlantooccipital and atlantoaxial levels, are associated with risk of injury to nearby structures including the dorsal root ganglion and the vertebral artery. A constellation of physical examination findings may aid in diagnosis abating the need for invasive anesthetic blocks. A suggested pattern of musculoskeletal signs discussed by Jull G et al. includes tenderness of upper cervical joints on palpation, reduced degree of cervical extension, and impaired cervical muscle function based on the Craniocervical Flexion Test (CCFT) [37]. Such a pattern of musculoskeletal signs may be more effective in the diagnosis of cervicogenic headache compared with a singular sign of cervical dysfunction [38]. While there are no specific radiologic abnormalities considered to be pathognomonic for cervicogenic headache, imaging modalities may help identify underlying cervical pathology such as disc degeneration and facet arthropathy. Conditions that commonly correspond with cervicogenic pain include cervical facet degeneration, soft tissue and facet injury resulting from cervical trauma, and postlaminectomy pain. Notably, when cervicogenic headache emerges as the predominant symptom postwhiplash, pain originating from the C2–C3 facet joint (third occipital neuralgia) often contributes to the headache presentation [39,40].

Patient selection criteria: Medial branch block

For diagnosing facet joint pain, it is recommended to conduct double controlled Medial Branch Blocks (MBBs) using two different local anesthetics with distinct onset and duration characteristics. The purpose of employing double controlled blocks is to minimize the risk of false positive results [41,42]. A positive response is defined as 80%–100% pain relief at symptomatic levels, aligning with the onset and duration of the specific local anesthetic used [42,43]. The expectation is that lidocaine, a short acting anesthetic, will yield a shorter analgesic effect compared with bupivacaine, which has a relatively longer duration of action [42,43]. Cutoffs for percentage pain relief, distinguishing positive from negative responses, vary from 50% to 100% [42,44,45]. While triple controlled MBBs with short and long acting anesthetics and saline injections are suggested to exclude placebo responses, they are deemed impractical and have not become routine practice [41–46].

To determine eligibility, an in depth diagnostic process is crucial, encompassing a comprehensive clinical history, physical examination, and pertinent imaging modalities such as X-ray, computed tomography (CT), and magnetic resonance imaging (MRI) [17,47–54]. Additionally, electrodiagnostic testing may be employed as needed. While these investigations aid in excluding other causes, a singular clinical marker or imaging finding is not considered sufficient to assuredly diagnose cervical facet mediated headache pain [54–56].

A definitive diagnosis of cervical facet mediated pain often hinges on the use of local anesthetic blocks, such as intraarticular facet joint injections or nerve blocks targeting the facet joint's nerve supply. To enhance diagnostic accuracy, physicians often conduct a confirmatory secondary block following the initial local anesthetic block before proceeding with cervical facet joint RFA [44,57–62].

TECHNIQUES FOR CERVICAL NERVE RADIOFREQUENCY ABLATION

Overview

The basic concept of radiofrequency ablation involves the application of radiofrequency currents through electrodes positioned near nociceptive pathways to interrupt the pain impulses. The thermal energy created by this technique generates a contained area of tissue destruction in regions that are transmitting the pain sensation. This procedure entails the precise application of radiofrequency waves to selectively target and destroy nerve tissue responsible for transmitting afferent pain signals. RFA procedures predominantly employ fluoroscopic guidance, complemented at times by CT or ultrasound, to ensure the accurate placement of insulated needles at specific nerve locations for optimal electrode insertion [1]. In the realm of cervical RFA, three primary modalities exist: thermal RFA (also known as conventional or continuous RFA), pulsed RFA, and water cooled RFA, each distinguished by their intricate technical parameters and procedural intricacies.

Radiofrequency ablation modalities

Thermal RFA, also referred to as conventional or continuous RFA, involves application of high frequency current (500 kHz) to generate high temperatures (60–90°C) for 90–120 s, leading to ablation of the target tissue through coagulation necrosis [63,64]. The size of the lesion depends on the size of the electrode, tissue temperature and the ablation duration. Lesions created by continuous RFA are well defined due to their circumscribed nature near the cannula tip.

Pulsed RFA involves the application of high frequency current in short intermittent pulses to modulate pain signals from sensory nerves without causing complete tissue destruction [65]. Uniquely, pulsed RFA is theorized to target A delta and C fibers, while leaving the nerve grossly intact [52]. A pulsed radiofrequency current is usually short (20 ms) with a high voltage burst followed by a silent phase (480 ms), with an oscillating frequency of 420 kHz during the pulse [66,67]. Due to the silent phases between pulses, heat is better able to dissipate, keeping the ablation temperature below 42°C, and minimizing complications such as neuritis and motor dysfunction [68,69].

Water cooled RFA is a recent iteration of the RFA technology and has been proposed to have a higher safety profile and improved long term efficacy in multiple spine mediated pain conditions.([19,21] Relative to pulsed and thermal RFA, water cooled RFA creates larger local neuronal lesions, which may increase the likelihood of a successful treatment, particularly in locations where nerve location can vary due to physiologic variability [70]. Because traditional RFA relies on heated probes at high temperatures, they may be increased risk of damage to adjacent tissue. With water cooled RFA, water circulates around the probes, allowing internally cooled probes to operate at 60°C, theoretically allowing more energy transfer to peripheral tissues due to reduced heat. This results in larger, deeper lesions leading to overall longer lasting pain relief [71].

PERTINENT ANATOMY

Radiofrequency ablation targets facet joints, which are synovial joints formed by superior and inferior articular processes and guide segmental motion. Cervical spinal nerves, much like other spinal regions, emerge through intervertebral foramina, except C1, which exits between the occipital bone and atlas (C1) [60]. Cervical nerves are numbered by the vertebra below, except for C8, which is below C7 and above T1 [60]. The cervical facet joints, extending from C2–3 to C7–T1, receive intricate innervation from the medial branches originating in the dorsal rami of cervical spinal nerve roots. Dorsal ramus divisions, both lateral and medial, innervate various muscles and provide cutaneous innervation [42]. This neural network includes specific branches such as the third occipital nerve (TON), which supplies the C2–C3 joint, the deep C3 medial branch linked to the C3–C4 joint, and the C4 to C8 medial branches associated with C3–C4 to C7–T1 facet joints. Notably, each facet joint in the cervical spine (C4–C8) receives innervation from two vertically adjacent spinal medial branches: the upper half is innervated by the medial branch from the upper level, and the lower half is innervated by the medial branch in the lower level [72].

In addition to the specific cervical joints, other anatomical structures and conditions must be considered prior to performing RFA. The spinal cord, housed within the vertebral canal formed by the stacked cervical vertebrae, stands as a critical bundle of nerve fibers and serves as the primary conduit for communication between the brain and the peripheral nervous system [73]. While oblique and sagittal needle insertions can be done to increase the area of thermal lesion along two separate sites, inappropriate needle trajectories and insertion depth can result in spinal cord injury [1]. Edema, characterized by the accumulation of excess fluid in the tissues, can be a concern postintervention, necessitating vigilant monitoring and appropriate management to mitigate its impact on both neural structures and surrounding soft tissues [74].

Muscles and soft tissues, including the cervical paraspinal muscles such as the splenius capitis, semispinalis capitis, and levator scapulae, play a multifaceted role in maintaining the intricate balance of the cervical spine. These muscles, with their unique anatomical orientations, contribute not only to stability but also to the dynamic functions of the neck, allowing for precise movements and postural adjustments. Ligaments and tendons further contribute to the structural integrity of the cervical spine, and bind bone to bone and muscle to bone, respectively. Inappropriate needle positioning can disrupt this intricate balance, further harming the relationship between each ligament, tendon, and muscle [73].

Arteries play a vital role in sustaining cerebral perfusion, and any inadvertent damage or disruption during the procedure could have serious consequences [73]. The vertebral arteries, ascending through the transverse foramina of the cervical vertebrae, are integral components of the vascular supply to the brain. Similarly, the carotid arteries, which are positioned along the anterior aspect of the cervical spine, supply blood to the head and face. Conditions such as atherosclerosis or tortuosity of blood vessels may pose additional challenges, emphasizing the need for thorough preprocedural assessments and imaging studies. Damage to this vasculature can result in a hemorrhagic or ischemic stroke.

RECOMMENDED APPROACHES

Currently there is no consensus on the best method for cervical medial branch blocks, as safety practices emphasize the treating physician's discretion in balancing the risks and benefits for each patient's unique situation. In general, a lateral approach to the third occipital nerve and C3 to C7 medial branch nerves is considered optimal, while a posterior approach to the C8 medial branch nerve is considered optimal due to the easier access to the target injection site with minimal tissue traversal, which can minimize procedure time and increase patient comfort [75,76]. The treating physician's comfort with the chosen approach may correlate with better outcomes. However, these considerations are mainly theoretical with limited literature comparing different approaches and exceptions may arise based on unique anatomical characteristics. Furthermore, while some research has explored the feasibility of ultrasound guided cervical medial branch blocks, the safety and diagnostic characteristics are still unclear, and fluoroscopic guidance remains the standard imaging modality [77–81].

Safety

Potential complications from cervical RFA, although rare, include infection, hemorrhage, numbness, dysesthesias, augmented pain at the procedural locus, and the phenomenon of deafferentation [7]. Additionally, occurrence of dysesthesia, characterized by a sensation akin to "sunburn", has been observed after cervical RFA; however, such occurrences tend to resolve spontaneously [82]. While the prospect of cervical nerve root impairment exists, adopting a secure technique guided by fluoroscopy and preceded by sensory and motor evaluations before lesioning can effectively mitigate this risk [7].

Efficacy

The current efficacy of radiofrequency ablation for headache treatment, specifically for treatment of cervicogenic headache is a subject of substantial investigation. Dysfunction within the cervical spine, primarily innervated by the C1–C3 spinal nerves, can lead to a cervicogenic headache, characterized by unilateral pain. Management of a cervicogenic headache poses challenges due to its overlap with migraine headache and limited diagnostic criteria. RFA of the cervical medial branch and third occipital nerve has emerged as a promising interventional therapy [83]. This technique aims to disrupt the afferent nerve supply responsible for CHA discomfort and has demonstrated effectiveness in treating chronic neck pain conditions [39,40]. However, it's noted that while RFA can alleviate pain, its effects are often transient, typically lasting until nerve regeneration or healing takes place [40,83,84]. Consequently, while RFA can help address the immediate pain symptoms associated with CHA, it does not necessarily address the underlying causes of the headache itself, which are often complex and not fully understood [60,61].

Studies demonstrate an average of 84% of patients achieve successful repeat radiofrequency ablation following an initial effective RFA procedure [47,48,50,85]. Within this subset of patients, the duration of beneficial effects span between 7 and 20 months. Notably, this mean slightly diverges from the outcomes reported in the systematic review and meta-analysis conducted by Smuck et al., which indicated a nonweighted average success rate of 88% (with a range of 67%–95%) for cases where the first RFA was successful [86]. The outcomes of denervation of the C2–C3 facet joint appear comparable with those observed with lower cervical levels, with approximately 77%–84% of patients exhibiting a favorable response (≥50% pain relief) after repeat RFA and maintaining this relief for durations ranging from 7.2 to 7.9 months [53,87].

Limitations of CRFA

Further large scale randomized controlled trials (RCTs) are needed to develop a consensus regarding the effectiveness of cervical nerve RFA for the treatment of cervicogenic headache. While this procedure can help alleviate pain from facet joint pathology, it may not address other pain generators including intervertebral discs and cervical musculature leading to incomplete pain relief in some patients [88].Clinical outcomes may be influenced by a variety of factors including operator skill, RFA techniques, and distribution of ablation. Previous systematic reviews have stated limited benefit of conventional or pulsed RFA in the short term [3,89]. Suer et al. stated Level II evidence (moderate to substantial benefit) based on the American Society of Interventional Pain Physicians (ASIPP) grading of evidence for the use of cervical nerve RFA for the management of cervical facet joint pain and cervicogenic headache. The review included three RCTs and reported postprocedure pain relief ranging from 30% to 50%. However, this review noted the included RCTs were of small sample size and only one study used diagnostic anesthetic blocks to establish facet joint pain. The included studies were heterogeneous regarding patient population, inclusion criteria, outcomes measures, and follow up interval.

As the origin of pain in cervicogenic headache has an anatomical basis at the cervical level, several procedures aim to reduce nociceptive input in the cervical region. A previous RCT found improved pain control at 9 months in patients undergoing pulsed RFA compared with greater occipital nerve block while another study reported no difference between these treatment modalities over a 12 month follow up period [90,91]. Additional information is needed regarding the comparative effectiveness and long term outcomes of available therapeutic options. Furthermore, studies comparing efficacy of different RFA modalities (thermal, pulsed, and water cooled RFA) for chronic neck pain are lacking. Future high powered studies should include a sham control group, validated outcome measures for pain and function, assessment of cervical mobility, as well as information about analgesic use following the procedure.

REFERENCES

[1] D.W. Lee, S. Pritzlaff, M.J. Jung, P. Ghosh, J.M. Hagedorn, J. Tate, K. Scarfo, N. Strand, K. Chakravarthy, D. Sayed, T.R. Deer, K. Amirdelfan, Latest evidence-based application for radiofrequency neurotomy (Learn): best practice guidelines from the american society of pain and neuroscience (aspn), Dove Medical Press Ltd, United States Journal of Pain Research 14 (2021) 2807–2831, https://doi.org/10.2147/JPR.S325665.

[2] T.A. Fredriksen, F. Antonaci, O. Sjaastad, Cervicogenic headache: too important to be left un-diagnosed, Springer-Verlag Italia s.r.l., Norway Journal of Headache and Pain 16 (1) (2015), https://doi.org/10.1186/1129-2377-16-6. Available from: https://thejournalofheadacheandpain.biomedcentral.com/.
[3] R.K. Grandhi, A.D. Kaye, A. Abd-Elsayed, Systematic review of radiofrequency ablation and pulsed radiofrequency for management of cervicogenic headaches, Curr. Pain Headache Rep. 22 (3) (2018), https://doi.org/10.1007/s11916-018-0673-9.
[4] C.A. Odonkor, T. Tang, D. Taftian, A. Chhatre, Bilateral intra-articular radiofrequency ablation for cervicogenic headache, Case Reports in Anesthesiology 2017 (2017) 1–6, https://doi.org/10.1155/2017/1483279.
[5] S. Verma, M. Tripathi, P. Chandra, Cervicogenic headache: current perspectives, Neurol 69 (7) (2021) S194, https://doi.org/10.4103/0028-3886.315992.
[6] J. Olesen, Headache classification committee of the international headache society (IHS) the international classification of headache disorders, 3rd edition, Cephalalgia 38 (1) (2018) 1–211, https://doi.org/10.1177/0333102417738202.
[7] J.M. Hagedorn, S. Golovac, T.R. Deer, N. Azeem, History and Development of Radiofrequency Ablation for Chronic Pain, Springer Science and Business Media LLC, 2021, pp. 3–6, https://doi.org/10.1007/978-3-030-78032-6_1.
[8] S. Aronow, The use of radio-frequency power in making lesions in the brain, J. Neurosurg. 17 (3) (1960) 431–438, https://doi.org/10.3171/jns.1960.17.3.0431.
[9] R.P. Pawl, Results in the treatment of low back syndrome from sensory neurolysis of the lumbar facets (facet rhizotomy) by thermal coagulation, Proc. Inst. Med. Chicago 30 (4) (1974) 151–152.
[10] t Banerjee, H.H. Pittman, Another armamentarium for treatment of low backache, N. C. Med. J. (7) (1976) 1976.
[11] J.A. McCulloch, L.W. Organ, Percutaneous radiofrequency lumbar rhizolysis (rhizotomy), Canadian Medical Association Journal 116 (1) (1977) 30–32.
[12] G. Flórez, J. Eiras, S. Ucar, Percutaneous rhizotomy of the articular nerve of luschka for low back and sciatic pain . Springer-Verlag Wien, Spain Acta Neurochirurgica, Supplementum 24 (1977) 67–71, https://doi.org/10.1007/978-3-7091-8482-0_11.
[13] R.J. Stolker, A.C.M. Vervest, L.M.P. Ramos, G.J. Groen, Electrode positioning in thoracic percutaneous partial rhizotomy: an anatomical study, Pain 57 (2) (1994) 241–251, https://doi.org/10.1016/0304-3959(94)90229-1.
[14] R.J. Stolker, A.C.M. Vervest, G.J. Groen, The treatment of chronic thoracic segmental pain by radiofrequency percutaneous partial rhizotomy, American Association of Neurological Surgeons, Netherlands Journal of Neurosurgery 80 (6) (1994) 986–992, https://doi.org/10.3171/jns.1994.80.6.0986.
[15] M. Van Kleef, G.A.M. Barendse, W.A.A.M. Dingemans, C. Wingen, R. Lousberg, S. De Lange, M.E. Sluijter, Effects of producing a radiofrequency lesion adjacent to the dorsal root ganglion in patients with thoracic segmental pain, Lippincott Williams and Wilkins, Netherlands Clinical Journal of Pain 11 (4) (1995) 325–332, https://doi.org/10.1097/00002508-199512000-00010.
[16] C.N. Shealy, Percutaneous radiofrequency denervation of spinal facets. Treatment for chronic back pain and sciatica, J. Neurosurg. 43 (4) (1975) 448–451, https://doi.org/10.3171/jns.1975.43.4.0448.
[17] S.M. Lord, L. Barnsley, N. Bogduk, Percutaneous radiofrequency neurotomy in the treatment of cervical zygapophysial joint pain: a caution, Neurosurgery 36 (4) (1995) 732–739, https://doi.org/10.1227/00006123-199504000-00014.
[18] S.P. Cohen, R.W. Hurley, C.C. Buckenmaier, C. Kurihara, B. Morlando, A. Dragovich, Randomized placebo-controlled study evaluating lateral branch radiofrequency denervation for sacroiliac joint pain, Lippincott Williams and Wilkins, United States, Anesthesiology 109 (2) (2008) 279–288, https://doi.org/10.1097/ALN.0b013e31817f4c7c.
[19] L. Kapural, F. Nageeb, M. Kapural, J.P. Cata, S. Narouze, N. Mekhail, Cooled radiofrequency system for the treatment of chronic pain from sacroiliitis: the first case-series, Pain Pract. 8 (5) (2008) 348–354, https://doi.org/10.1111/j.1533-2500.2008.00231.x.

[20] L. Kapural, A. Ng, J. Dalton, E. Mascha, M. Kapural, M.de La Garza, M. Nagy, Intervertebral disc biacuplasty for the treatment of lumbar discogenic pain: results of a six-month follow-up, Pain Med. 9 (1) (2008) 60–67, https://doi.org/10.1111/j.1526-4637.2007.00407.x.

[21] L Kapural, B Vrooman, S Sarwar, Radiofrequency intradiscal biacuplasty for treatment of discogenic lower back pain: a 12-month follow-up. doi: 10.1111/pme.12266.

[22] M.E. Sluijter, E.R. Cosman, W.B. Rittman, M. Van Kleef, The effects of pulsed radiofrequency fields applied to the dorsal root ganglion: a preliminary report, Pain Clin. 11 (2) (1998) 109–117.

[23] D.M. Long, M. BenDebba, W.S. Torgerson, R.J. Boyd, E.G. Dawson, R.W. Hardy, J.T. Robertson, G.W. Sypert, C. Watts, Persistent back pain and sciatica in the United States: patient characteristics, J. Spinal Disord. 9 (1) (1996) 40–58.

[24] J.F. Hamer, T.A. Purath, Response of cervicogenic headaches and occipital neuralgia to radiofrequency ablation of the C2 dorsal root ganglion and/or third occipital nerve, Headache J. Head Face Pain 54 (3) (2014) 500–510, https://doi.org/10.1111/head.12295.

[25] W. Halim, N.H.L. Chua, K.C. Vissers, Long-term pain relief in patients with cervicogenic headaches after pulsed radiofrequency application into the lateral atlantoaxial (C1-2) joint using an anterolateral approach, Pain Pract. 10 (4) (2010) 267–271, https://doi.org/10.1111/j.1533-2500.2010.00360.x.

[26] J.-Q. Hu, J. Zhang, B. Ru, W.-J. Cai, W.-L. Liu, R. Guo, Z.-W. Ji, Q. Wan, L.-H. Xu, Y. Cheng, J. Zhang, S. Li, Computed tomography -guided radiofrequency ablation of cervical intervertebral discs for the treatment of refractory cervicogenic headache: a retrospective chart review, Headache J. Head Face Pain 62 (7) (2022) 839–847, https://doi.org/10.1111/head.14361.

[27] Narouze, Chapter 22 - Cervicogenic Headache Essentials of Pain Medicine, Elsevier, 2018, pp. 177–182.

[28] R. Castien, W. De Hertogh, A neuroscience perspective of physical treatment of headache and neck pain, Front. Neurol. 10 (2019), https://doi.org/10.3389/fneur.2019.00276.

[29] Y. Khalili, L. Al, P.B. Murphy, Cervicogenic headache, StatPearls (2022) 2022.

[30] D. Swanson, R. Guedry, M. Boudreaux, E. Muhlenhaupt, A.D. Kaye, O. Viswanath, I. Urits, An update on the diagnosis, treatment, and management of occipital neuralgia, J. Craniofac. Surg. 33 (3) (2022) 779–783, https://doi.org/10.1097/scs.0000000000008360.

[31] S. Goyal, A. Kumar, P. Mishra, D. Goyal, Efficacy of interventional treatment strategies for managing patients with cervicogenic headache: a systematic review, Korean Journal of Anesthesiology 75 (1) (2022) 12–24, https://doi.org/10.4097/kja.21328.

[32] J.H. Talbet, A.G. Elnahry, OnabotulinumtoxinA for the treatment of headache: an updated review, IMR Press Limited, United States Journal of Integrative Neuroscience 21 (1) (2022), https://doi.org/10.31083/j.jin2101037. Available from: https://imrpress.com/journal/JIN/21/1/10.31083/j.jin2101037.

[33] E.C. Rodriguez-Merchan, A.D. Delgado-Martinez, J.D. Andres-Ares, Radiofrequency ablation for the management of pain of spinal origin in orthopedics, Mashhad University of Medical Sciences, Spain, Archives of Bone and Joint Surgery 11 (11) (2023) 666–671, https://doi.org/10.22038/ABJS.2023.71327.3333.

[34] H. Benzon, S. Fishman, S. Liu, S.P. Cohen, S.N. Raja, Essentials of Pain Medicine Essentials of Pain Medicine, Elsevier Inc., United States Elsevier Inc., United States, 2011, https://doi.org/10.1016/C2009-0-61078-3. Available from: http://www.sciencedirect.com/science/book/9781437722420.

[35] H. Göbel, Classification of headaches, Cephalalgia 21 (7) (2001) 770–773, https://doi.org/10.1046/j.0333-1024.2001.00007.x.

[36] O. Sjaastad, T.A. Fredriksen, V. Pfaffenrath, Cervicogenic headache: diagnostic criteria, Headache J. Head Face Pain 30 (11) (1990) 725–726, https://doi.org/10.1111/j.1526-4610.1990.hed3011725.x.

[37] G. Jull, Cervicogenic headache, Musculoskeletal Science and Practice 66 (2023) 102787, https://doi.org/10.1016/j.msksp.2023.102787.

[38] S.L. Getsoian, S.M. Gulati, I. Okpareke, R.J. Nee, G.A. Jull, Validation of a clinical examination to differentiate a cervicogenic source of headache: a diagnostic prediction model using controlled diagnostic blocks, BMJ Publishing Group, United States BMJ Open 10 (5) (2020), https://doi.org/10.1136/bmjopen-2019-035245. Available from: http://bmjopen.bmj.com/content/early/by/section.
[39] Hogg-Johnson, Velde, Carroll, The burden and determinants of neck pain in the general population: results of the bone and joint decade 2000-2010 task force on neck pain and its associated disorders, Spine (Phila Pa) (1976), https://doi.org/10.1097/BRS.0B013E31816454C81976.
[40] S.M. Lord, L. Barnsley, B.J. Wallis, N. Bogduk, Chronic cervical zygapophysial joint pain after whiplash: a placebo- controlled prevalence study, Spine 21 (15) (1996) 1737−1745, https://doi.org/10.1097/00007632-199608010-00005.
[41] Les Barnsley, S. Lord, B. Wallis, N. Bogduk, False-positive rates of cervical zygapophysial joint blocks, Clin. J. Pain 9 (2) (1993) 124−130, https://doi.org/10.1097/00002508-199306000-00007.
[42] B. Nikolai, M.G. Brian, Management of Acute and Chronic Neck Pain : An Evidence-Based Approach, 2006, p. 2006.
[43] Les Barnsley, S. Lord, N. Bogduk, Comparative local anaesthetic blocks in the diagnosis of cervical zygapophysial joint pain, Pain 55 (1) (1993) 99−106, https://doi.org/10.1016/0304-3959(93)90189-v.
[44] S.C. Holz, What is the correlation between facet joint radiofrequency outcome and response to comparative medial branch blocks? Pain Physician 19 (3) (2016) 163−172, https://doi.org/10.36076/ppj/2019.19.163.
[45] A. Engel, W. King, B.J. Schneider, B. Duszynski, N. Bogduk, The effectiveness of cervical medial branch thermal radiofrequency neurotomy stratified by Selection Criteria: a systematic review of the Literature, Oxford University Press, United States, Pain Med. 21 (11) (2020) 2726−2737, https://doi.org/10.1093/PM/PNAA219.
[46] S.P. Cohen, W.M. Hooten, Advances in the diagnosis and management of neck pain, BMJ Publishing Group, United States, BMJ 358 (2017), https://doi.org/10.1136/bmj.j3221. Available from: http://www.bmj.com/.
[47] G.J. McDonald, S.M. Lord, N. Bogduk, Long-term follow-up of patients treated with cervical radiofrequency neurotomy for chronic neck pain, Lippincott Williams and Wilkins, Australia, Neurosurgery 45 (1) (1999) 61−68, https://doi.org/10.1097/00006123-199907000-00015. Available from:.
[48] Les Barnsley, Percutaneous radiofrequency neurotomy for chronic neck pain: outcomes in a series of consecutive patients, Pain Med. 6 (4) (2005) 282−286, https://doi.org/10.1111/j.1526-4637.2005.00047.x.
[49] D.A. Sapir, J.M. Gorup, Radiofrequency medial branch neurotomy in litigant and non-litigant patients with cervical whiplash, Spine 26 (12) (2001) e268, https://doi.org/10.1097/00007632-200106150-00016.
[50] S.P. Cohen, Z.H. Bajwa, J.J. Kraemer, A. Dragovich, K.A. Williams, J. Stream, A. Sireci, G. McKnight, R.W. Hurley, Factors predicting success and failure for cervical facet radiofrequency denervation: a multi-center analysis, Reg. Anesth. Pain Med. 32 (6) (2007) 495−503, https://doi.org/10.1016/j.rapm.2007.05.009.
[51] J. Macvicar, J.M. Borowczyk, A.M. Macvicar, B.M. Loughnan, N. Bogduk, Cervical medial branch radiofrequency neurotomy in New Zealand, Blackwell Publishing Inc., New Zealand, Pain Med. 13 (5) (2012) 647−654, https://doi.org/10.1111/j.1526-4637.2012.01351.x.
[52] G.C. Speldewinde, Outcomes of percutaneous zygapophysial and sacroiliac joint neurotomy in a community setting, Blackwell Publishing Inc., Australia Pain Medicine 12 (2) (2011) 209−218, https://doi.org/10.1111/j.1526-4637.2010.01022.x.
[53] J. Govind, W. King, B. Bailey, N. Bogduk, Radiofrequency neurotomy for the treatment of third occipital headache, J. Neurol. Neurosurg. Psychiatry 74 (1) (2003) 88−93, https://doi.org/10.1136/jnnp.74.1.88.

[54] N. Bogduk, P. Dreyfuss, R. Baker, W. Yin, M. Landers, M. Hammer, C. Aprill, Complications of spinal diagnostic and treatment procedures, Pain Med. 9 (Suppl. 1) (2008) S11, https://doi.org/10.1111/j.1526-4637.2008.00437.x.
[55] O. Palea, H.M. Andar, R. Lugo, M. Granville, R.E. Jacobson, Direct posterior bipolar cervical facet radiofrequency rhizotomy: a simpler and safer approach to denervate the facet capsule, Cureus (2018), https://doi.org/10.7759/cureus.2322.
[56] M. van Eerd, A. Lataster, M. Sommer, P. Jacob, M. van Kleef, A modified posterolateral approach for radiofrequency denervation of the medial branch of the cervical segmental nerve in cervical facet joint pain based on anatomical considerations, Pain Pract. 17 (5) (2017) 596−603, https://doi.org/10.1111/papr.12499.
[57] S.P. Cohen, S.A. Strassels, C. Kurihara, S.R. Griffith, B. Goff, K. Guthmiller, H.T. Hoang, B. Morlando, C. Nguyen, Establishing an optimal "cutoff" threshold for diagnostic lumbar facet blocks: a prospective correlational study, Clin. J. Pain 29 (5) (2013) 382−391, https://doi.org/10.1097/AJP.0b013e31825f53bf.
[58] R. Derby, I. Melnik, J.E. Lee, S.H. Lee, Correlation of lumbar medial branch neurotomy results with diagnostic medial branch block cutoff values to optimize therapeutic outcome, Blackwell Publishing Inc., United States, Pain Med. 13 (12) (2012) 1533−1546, https://doi.org/10.1111/j.1526-4637.2012.01500.x.
[59] S.P. Cohen, K.A. Williams, C. Kurihara, C. Nguyen, C. Shields, P. Kim, S.R. Griffith, T.M. Larkin, M. Crooks, N. Williams, B. Morlando, S.A. Strassels, Multicenter, randomized, comparative cost-effectiveness study comparing 0, 1, and 2 diagnostic medial branch (Facet Joint Nerve) block treatment paradigms before lumbar facet radiofrequency denervation, Lippincott Williams and Wilkins, United States, Anesthesiology 113 (2) (2010) 395−405, https://doi.org/10.1097/ALN.0b013e3181e33ae5.
[60] T. Kweon, J. Kim, H. Lee, M. Kim, Y.W. Lee, Anatomical analysis of medial branches of dorsal rami of cervical nerves for radiofrequency thermocoagulation, Lippincott Williams and Wilkins, South Korea Regional Anesthesia and Pain Medicine 39 (6) (2014) 465−471, https://doi.org/10.1097/AAP.0000000000000175.
[61] L. Manchikanti, J.A. Hirsch, A.D. Kaye, M.V. Boswell, Cervical zygapophysial (facet) joint pain: effectiveness of interventional management strategies, Taylor and Francis Inc., United States, PGM (Postgrad. Med.) 128 (1) (2016) 54−68, https://doi.org/10.1080/00325481.2016.1105092.
[62] L. Manchikanti, Comprehensive evidence-based guidelines for facet joint interventions in the management of chronic spinal pain . ASIPP) guidelines, ASIPP) guidelines.
[63] H. Shanthanna, P. Chan, J. McChesney, J. Paul, L. Thabane, Assessing the effectiveness of 'pulse radiofrequency treatment of dorsal root ganglion' in patients with chronic lumbar radicular pain: study protocol for a randomized control trial, Trials 13 (1) (2012), https://doi.org/10.1186/1745-6215-13-52.
[64] D. Byrd, S. Mackey, Pulsed radiofrequency for chronic pain, Curr. Pain Headache Rep. 12 (1) (2008) 37−41, https://doi.org/10.1007/s11916-008-0008-3.
[65] E.J. Choi, Y.M. Choi, E.J. Jang, J.Y. Kim, T.K. Kim, K.H. Kim, Neural ablation and regeneration in pain practice, Korean Pain Society, South Korea Korean Journal of Pain 29 (1) (2016) 3−11, https://doi.org/10.3344/kjp.2016.29.1.3.
[66] J. Mata, P. Valentí, B. Hernández, B. Mir, J.L. Aguilar, Study protocol for a randomised controlled trial of ultrasound-guided pulsed radiofrequency of the genicular nerves in the treatment of patients with osteoarthritis knee pain, BMJ Publishing Group, Spain BMJ Open 7 (11) (2017), https://doi.org/10.1136/bmjopen-2017-016377. Available from: http://bmjopen.bmj.com/content/early/by/section.
[67] Y. Ding, H. Li, T. Hong, R. Zhao, P. Yao, G. Zhao, Efficacy and safety of computed tomography-guided pulsed radiofrequency modulation of thoracic dorsal root ganglion on herpes zoster neuralgia, Blackwell Publishing Inc., China, Neuromodulation 22 (1) (2019) 108−114, https://doi.org/10.1111/ner.12858. Available from: http://onlinelibrary.wiley.com/journal/10.1111/(ISSN)1525-1403.
[68] K. Tun, B. Cemil, A.G. Gurcay, E. Kaptanoglu, M.F. Sargon, I. Tekdemir, A. Comert, Y. Kanpolat, Ultrastructural evaluation of pulsed radiofrequency and conventional

radiofrequency lesions in rat sciatic nerve, Surg. Neurol. 72 (5) (2009) 496–500, https://doi.org/10.1016/j.surneu.2008.11.016.
[69] Z. Abbott, M. Smuck, A. Haig, O. Sagher, Irreversible spinal nerve injury from dorsal ramus radiofrequency neurotomy: a case report, Arch. Phys. Med. Rehabil. 88 (10) (2007) 1350–1352, https://doi.org/10.1016/j.apmr.2007.07.006.
[70] K. Malik, H.T. Benzon, D. Walega, Water-cooled radiofrequency: a neuroablative or a neuromodulatory modality with broader applications? Case Reports in Anesthesiology 2011 (2011) 1–3, https://doi.org/10.1155/2011/263101.
[71] L.O. Oladeji, J.L. Cook, Cooled radio frequency ablation for the treatment of osteoarthritis-related knee pain: evidence, indications, and outcomes, Georg Thieme Verlag, United States Journal of Knee Surgery 32 (1) (2019) 65–71, https://doi.org/10.1055/s-0038-1675418.
[72] M.A. Huntoon, Anatomy of the cervical intervertebral foramina: vulnerable arteries and ischemic neurologic injuries after transforaminal epidural injections, Pain 117 (1–2) (2005) 104–111, https://doi.org/10.1016/j.pain.2005.05.030.
[73] R.W. Hurley, M.C.B. Adams, M. Barad, A. Bhaskar, A. Bhatia, A. Chadwick, T.R. Deer, J. Hah, W.M. Hooten, N.R. Kissoon, D.W. Lee, Z. McCormick, J.Y. Moon, S. Narouze, D.A. Provenzano, B.J. Schneider, M. van Eerd, J. Van Zundert, M.S. Wallace, S.M. Wilson, Z. Zhao, S.P. Cohen, Consensus practice guidelines on interventions for cervical spine (facet) joint pain from a multispecialty international working group, BMJ Publishing Group, United States Regional Anesthesia and Pain Medicine 47 (1) (2022) 3–59, https://doi.org/10.1136/rapm-2021-103031.
[74] M.T. Nevalainen, P.J. Foran, J.B. Roedl, A.C. Zoga, W.B. Morrison, Cervical facet oedema: prevalence, correlation to symptoms, and follow-up imaging, Clin. Radiol. 71 (6) (2016) 570–575, https://doi.org/10.1016/j.crad.2016.02.026.
[75] S.P. Cohen, S.A. Strassels, C. Kurihara, A. Forsythe, C.C. Buckenmaier, B. McLean, G. Riedy, S. Seltzer, Randomized study assessing the accuracy of cervical facet joint nerve (medial branch) blocks using different injectate volumes, Lippincott Williams and Wilkins, United States, Anesthesiology 112 (1) (2010) 144–152, https://doi.org/10.1097/ALN.0b013e3181c38a82.
[76] Y.I. Lee, H.J. Soh, E.D. Kim, Postdural puncture headache after cervical medial branch block, Soonchunhyang Medical Science 24 (2) (2018) 196–198, https://doi.org/10.15746/sms.18.037.
[77] R.J. Finlayson, J.P.B. Etheridge, L. Vieira, G. Gupta, D.Q.H. Tran, A randomized comparison between ultrasound- and fluoroscopy-guided third occipital nerve block, Reg. Anesth. Pain Med. 38 (3) (2013) 212–217, https://doi.org/10.1097/AAP.0b013e31828b25bc.
[78] A. Siegenthaler, S. Mlekusch, S. Trelle, J. Schliessbach, M. Curatolo, U. Eichenberger, Accuracy of ultrasound-guided nerve blocks of the cervical zygapophysial joints, Anesthesiology 117 (2) (2012) 347–352, https://doi.org/10.1097/aln.0b013e3182605e11.
[79] K.D. Park, D.J. Lim, W.Y. Lee, J.K. Ahn, Y. Park, Ultrasound versus fluoroscopy-guided cervical medial branch block for the treatment of chronic cervical facet joint pain: a retrospective comparative study, Springer Verlag, South Korea Skeletal Radiology 46 (1) (2017) 81–91, https://doi.org/10.1007/s00256-016-2516-2.
[80] R.J. Finlayson, G. Gupta, M. Alhujairi, S. Dugani, D.Q.H. Tran, Cervical medial branch block: a novel technique using ultrasound guidance, Reg. Anesth. Pain Med. 37 (2) (2012) 219–223, https://doi.org/10.1097/AAP.0b013e3182374e24.
[81] B. Schneider, A. Popescu, C. Smith, Ultrasound imaging for cervical injections, Oxford University Press, United States, Pain Med. 21 (1) (2020) 196–197, https://doi.org/10.1093/pm/pnz277.
[82] R. Reddy, S. Zardouz, S. Rejai, J. Chen, Hyperhidrosis and dysautonomia in a patient with a history of tetraplegia following cervical facet radiofrequency ablation, PM&R 11 (12) (2019) 1354–1356, https://doi.org/10.1002/pmrj.12168.
[83] L. Barnsley, S.M. Lord, B.J. Wallis, N. Bogduk, The prevalence of chronic cervical zygapophysial joint pain after whiplash, Spine 20 (1) (1995) 20–26, https://doi.org/10.1097/00007632-199501000-00004.

[84] M.B. Furman, Atlas of Image-Guided Spinal Procedures Atlas of Image-Guided Spinal Procedures, Elsevier, United States Elsevier, United States, 2017, pp. 1–650, https://doi.org/10.1016/B978-0-323-40153-1.27001-7. Available from: https://www.sciencedirect.com/book/9780323401531.

[85] D.S. Husted, D. Orton, J. Schofferman, K. Garrett, Effectiveness of repeated radiofrequency neurotomy for cervical facet joint pain, J. Spinal Disord. Tech. 21 (6) (2008) 406–408, https://doi.org/10.1097/bsd.0b013e318158971f.

[86] M. Smuck, R.A. Crisostomo, K. Trivedi, D. Agrawal, Success of initial and repeated medial branch neurotomy for zygapophysial joint pain: a systematic review, PM&R 4 (9) (2012) 686–692, https://doi.org/10.1016/j.pmrj.2012.06.007.

[87] S.M. Lord, G.J. McDonald, N. Bogduk, Percutaneous radiofrequency neurotomy of the cervical medial branches: a validated treatment for cervical zygapophysial joint pain, Lippincott Williams and Wilkins, Australia, Neurosurg. Q. 8 (4) (1998) 288–308, https://doi.org/10.1097/00013414-199812000-00004.

[88] m Suer, S Wahezi, A Abd-Elsayed, N. Sehgal, Systematic review cervical facet joint pain and cervicogenic headache treated with radiofrequency ablation: a systematic review.

[89] C. Ekhator, A. Urbi, B.N. Nduma, S. Ambe, E. Fonkem, Safety and efficacy of radiofrequency ablation and epidural steroid injection for management of cervicogenic headaches and neck pain: meta-analysis and literature review, Cureus (2023), https://doi.org/10.7759/cureus.34932.

[90] T. Gabrhelík, P. Michálek, M. Adamus, Pulsed radiofrequency therapy versus greater occipital nerve block in the management of refractory cervicogenic headache - a pilot study, Prague Med. Rep. 112 (4) (2011) 279–287.

[91] S.R.S. Haspeslagh, H.A. Van Suijlekom, I.E. Lamé, A.G.H. Kessels, M. van Kleef, W.E.J. Weber, Randomised controlled trial of cervical radiofrequency lesions as a treatment for cervicogenic headache [ISRCTN07444684], BMC Anesthesiol. 6 (2006), https://doi.org/10.1186/1471-2253-6-1Netherlands. Available from: http://www.biomedcentral.com/1471-2253/6/1.

Chapter | Nine

Occipital nerve radiofrequency ablation

Philip M. Stephens[1], Richard W. Kim[2], Casey Brown[1]

[1]Department of Physical Medicine & Rehabilitation, University of Pittsburgh Medical Center, Pittsburgh, PA, United States; [2]Department of Physical Medicine & Rehabilitation, NYP Colombia Cornell, New York, NY, United States

INTRODUCTION

Occipital Nerve Radiofrequency Ablation (RFA) or Radiofrequency Neurotomy (RFN) continues to gain attention in the field of pain medicine as a potential treatment alternative for occipital neuralgia (ON), cluster headaches, migraines, and various cervicogenic headaches mediated by the occipital nerve refractory to less invasive procedures. Electrocautery first dates back to the 1920s with the development of a device now known as "the Bovie Knife" for the excision of hepatocellular lesions, followed by the first applications of neurotomy in cardiac arrhythmia ablations [1,2]. RFA was first used almost a century ago when German doctor, Martin Kirschner, targeted the Gasserian ganglion with radiofrequency currents to treat trigeminal neuralgia [3]. RFA works similarly to electrocautery through the application of alternating currents, delivering therapeutic disruption of the target, such as blood vessels or in the field of pain medicine— nerves. This minimally invasive procedure interrupts the transmission of pain signals to the brain through A-delta and C fiber by inducing a thermal lesion. RFA gained acceptance within pain medicine in the 1970s with initial applications to nerves facilitating spinal facet joint pain, followed by sacroiliac joint pain and end-stage knee osteoarthritis [4—7]. By the late 1990s, attention turned to applications mediated by the occipital nerve to treat ON, a condition characterized by the International Society of Headaches as severe pain of the posterior neck and scalp, which can lead to intense sensations of piercing, throbbing, and electrical stimuli within the distribution of the greater, lesser, and/or third occipital nerves [8,9]. RFA continues to gain traction outside the treatment of ON with expanding indications for occipital nerve RFA in the treatment of various headache disorders as the targets, indications, and technical expertise improve.

Interventional Management of Migraines and Other Headache Disorders. https://doi.org/10.1016/B978-0-443-23557-3.00009-7

INDICATIONS FOR OCCIPITAL NERVE RADIOFREQUENCY ABLATION

ON was the first condition mediated by the occipital nerve to be treated with RFA, now expanding into cervicogenic headaches, tension headaches, and migraines [10,11]. Patient selection, as with any other medical treatment, is critical to matching the patient with the best treatment to alleviate their pain. Before proceeding with occipital nerve RFA, the patient must fail conservative measures (i.e., medications, physical therapy, and lifestyle modifications), determination of significant impact on quality of life, and undergo a successful diagnostic occipital nerve block to confirm the pain generator [12]. Benefits of this procedure include minimal recovery time, low risk of harm, precise location under fluoroscopic guidance, and maximal potential for relief from symptoms following successful diagnostic blocks. Although this procedure holds great promise, there continues to be a limited selection of data to bolster its efficacy. Drawbacks include temporary discomfort during the procedure, the risk of damage to blood vessels, and the incomplete generalizability of this technique to other headache types.

Techniques for occipital nerve radiofrequency ablation

As with nearly all pain medicine procedures, a thorough understanding of the relevant anatomy is critical to performing a safe and effective procedure. The Occipital nerves—Greater, Lesser, and Third—arise from the spinal nerves of C2 and C3. These nerves' primary function is sensory innervation of the posterior and lateral scalp with the Third Occipital Nerve (TON) providing motor input to the semispinalis capitis. The Greater Occipital Nerve (GON) originates from the dorsal ramus of C2 before piecing the semispinalis capitis then running superiorly while medial to the occipital artery—serving as a landmark for ultrasound identification. The GON provides sensory innervation to the majority of the posterior scalp. Located most laterally of the three branches, the Lesser Occipital Nerve (LON) provides innervation to the lateral scalp and external ear from the ventral rami of C2—C3. Landmark guidance will locate the GON one-third distance from the occipital protuberance to the mastoid process, while the LON is two-thirds distance between the same structures. Lastly, the Third Occipital Nerve (TON) arises from the dorsal rami of C3 innervating the C2—C3 Facet before traveling most medially to supply sensation to the nuchal area and occipital protuberance. Physical exam utilizing Tinel's can be helpful to identify the problematic branch. It is important to note that communications of these branches may reach the fronto-orbital area through trigeminocervical interneuronal connections in the trigeminal spinal nuclei [8]. These types of headaches are most commonly triggered by compression of the GON 90% or LON 10% [13].

Based on the abovementioned anatomical location of the corresponding occipital nerve branch pain distribution, physical exam, and successful diagnostic block, a wheel of lidocaine 1% is injected prior to the insertion of the RFA needle. Ideally, the RFA needle would be placed as parallel as possible to the target nerve to improve electrode contact surface area to improve burning localization. Electrical stimulation is then completed at 50 Hz to ensure proximity to the target symptomatic branch with paresthesia in the appropriate distribution lower than 0.5V. Needle adjustments are made to influence the greatest stimulation at the lowest voltage. Goal impedance range is between 150 and 500 Ω to ensure appropriate proximity. Lidocaine 2% is injected with time given for effect onset before proceeding. Pulsed radiofrequency (PRF) ablation current, depending on manufacturing settings, is typically delivered in 20 ms pulses in a 1 second cycle for 120−240 seconds at 42°C [14,15]. Of note, another technique utilizes thermal radiofrequency ablation, delivering 80°C continuously for 180 s, though this is less often utilized than PRF due to patient discomfort [16]. Although a common practice, no significant difference has been demonstrated in ablation outcomes or pain scores with postprocedure steroid injection prior to needle removal [17].

Efficacy and safety of occipital nerve radiofrequency ablation

Several key studies have pushed this area of pain medicine forward regarding the treatment of ON, migraines, and other occipital nerve-mediated headaches. One early study, evaluating 102 patients diagnosed with ON treated with PRF, found 51% of participants experienced ≥50% pain relief and satisfaction with treatment lasting ≥3 months. Positive predictive variables for favorable treatment outcomes included traumatic inciting event, lower volumes of successful diagnostic blocks, and multiple rounds of PRF. Negative predictive variables included pain anterior to the scalp apex and ongoing secondary gain issues [14]. Another study compared PRF with local anesthetic and saline utilizing a randomized, double-blind design with 81 participants. Greater occipital pain reduction was measured at 6 weeks through 3 months following three 120 s cycles of PRF compared with steroid plus sham PRF [18]. Yet another study evaluated 277 patients retrospectively who underwent occipital nerve RFA with adequate follow-up, observing an average pain reduction of 63.5% with a mean duration of pain improvement of 254 days [19]. As for complications with RFA, most are self-limiting including dysesthesia, hypersensitivity, swelling, and temporary worsening of pain (<21 days) with complication rates independent of treatment success [14,18−20]. These findings highlight the potential relief RFA could provide from occipital nerve-mediated pain and provide enthusiasm for increased application within this patient population with an understanding of potential variable outcomes and typically temporary side effects.

REFERENCES

[1] A. Armaiz Flores, M.L. Oppenheimer Velez, S.M. Thompson, A.J. Windebank, A.J. Greenberg-Worisek, Navigating the clinical translation of medical devices: the case of radiofrequency ablation, Clin. Transl. Sci. 11 (1) (2018) 8−10, https://doi.org/10.1111/cts.12516.

[2] M. Habibi, R.D. Berger, H. Calkins, Radiofrequency ablation: technological trends, challenges, and opportunities, Europace 23 (4) (2021) 511−519, https://doi.org/10.1093/europace/euaa328.

[3] M. Krischner, Zur elektrochirurgie [for electrosurgery], Arch. Klin. Chir. 147 (1931) 761.

[4] C.N. Shealy, Facet denervation in the management of back and sciatic pain, Clin. Orthop. Relat. Res. 115 (1976) 157−164.

[5] R.P. Pawl, Results in the treatment of low back syndrome from sensory neurolysis of the lumbar facets (facet rhizotomy) by thermal coagulation, Proc. Inst. Med. Chic. 30 (4) (1974) 151−152.

[6] F.M. Ferrante, L.F. King, E.A. Roche, et al., Radiofrequency sacroiliac joint denervation for sacroiliac syndrome, Reg. Anesth. Pain Med. 26 (2) (2001) 137−142, https://doi.org/10.1053/rapm.2001.21739.

[7] W.J. Choi, S.J. Hwang, J.G. Song, et al., Radiofrequency treatment relieves chronic knee osteoarthritis pain: a double-blind randomized controlled trial, Pain 152 (3) (2011) 481−487, https://doi.org/10.1016/j.pain.2010.09.029.

[8] Headache classification committee of the international headache society (IHS) the international classification of headache disorders, 3rd edition, Cephalalgia 38 (1) (2018) 1−211, https://doi.org/10.1177/0333102417738202.

[9] M. Hammer, W. Meneese, Principles and practice of radiofrequency neurolysis, Curr. Rev. Pain 2 (4) (1998) 267−278.

[10] A. Abd-Elsayed, L. Kreuger, S. Wheeler, J. Robillard, S. Seeger, D. Dulli, Radiofrequency ablation of pericranial nerves for treating headache conditions: a promising option for patients, Ochsner J. 18 (1) (2018) 59−62.

[11] J.F. Hamer, T.A. Purath, Response of cervicogenic headaches and occipital neuralgia to radiofrequency ablation of the C2 dorsal root ganglion and/or third occipital nerve, Headache 54 (3) (2014) 500−510, https://doi.org/10.1111/head.12295.

[12] D. Swanson, R. Guedry, M. Boudreaux, et al., An update on the diagnosis, treatment, and management of occipital neuralgia, J. Craniofac. Surg. 33 (3) (2022) 779−783, https://doi.org/10.1097/SCS.0000000000008360.

[13] S.R. Hammond, G. Danta, Occipital neuralgia, Clin. Exp. Neurol. 15 (1978) 258−270.

[14] J.H. Huang, S.M. Galvagno Jr., M. Hameed, et al., Occipital nerve pulsed radiofrequency treatment: a multi-center study evaluating predictors of outcome, Pain Med. 13 (4) (2012) 489−497, https://doi.org/10.1111/j.1526-4637.2012.01348.x.

[15] P. Vanelderen, T. Rouwette, P. De Vooght, et al., Pulsed radiofrequency for the treatment of occipital neuralgia: a prospective study with 6 months of follow-up, Reg. Anesth. Pain Med. 35 (2) (2010) 148−151, https://doi.org/10.1097/aap.0b013e3181d24713.

[16] L.M. Hoffman, A. Abd-Elsayed, T.J. Burroughs, H. Sachdeva, Treatment of occipital neuralgia by thermal radiofrequency ablation, Ochsner J. 18 (3) (2018) 209−214, https://doi.org/10.31486/toj.17.0104.

[17] A. Abd-Elsayed, M. Loebertman, P. Huynh, I. Urits, O. Viswanath, N. Sehgal, The long-term efficacy of radiofrequency ablation with and without steroid injection, Psychopharmacol. Bull. 50 (4 Suppl. 1) (2020) 11−16.

[18] S.P. Cohen, B.L. Peterlin, L. Fulton, et al., Randomized, double-blind, comparative-effectiveness study comparing pulsed radiofrequency to steroid injections for occipital neuralgia or migraine with occipital nerve tenderness, Pain 156 (12) (2015) 2585−2594, https://doi.org/10.1097/j.pain.0000000000000373. PMID: 26447705; PMCID: PMC4697830.

[19] A. Abd-Elsayed, S.A. Yapo, N.N. Cao, M.K. Keith, K.J. Fiala, Radiofrequency ablation of the occipital nerves for treatment of neuralgias and headache, Pain Pract. 24 (1) (2024) 18−24, https://doi.org/10.1111/papr.13276. Epub 2023 Jul 17. PMID: 37461297.

[20] V. Orhurhu, L. Huang, R.C. Quispe, et al., Use of radiofrequency ablation for the management of headache: a systematic review, Pain Phys. 24 (7) (2021) E973−E987.

Chapter | Ten

Occipital nerve stimulation

Zachary Danssaert[1,2], Ricky Ju[3], Mihir Jani[4], Alan David Kaye[5]

[1]*Weill Cornell Department of Rehabilitation Medicine, New York, NY, United States;* [2]*Columbia University Department of Rehabilitation and Regenerative Medicine, New York, NY, United States;* [3]*Burke Hospital Department of Rehabilitation, White Plains, NY, United States;* [4]*Montefiore Medical Center Department of Rehabilitation, Bronx, NY, United States;* [5]*Department of Anesthesiology, Louisiana State University Health Sciences Center, Shreveport, LA, United States*

INTRODUCTION

History of occipital nerve stimulation

Occipital nerve stimulation (ONS) is a neuromodulation technique in headache management that was first introduced by Dr. Picaza, Dr. Hunter, and Dr. Cannon from the Department of Neurosurgery at the University of Tennessee in 1977. In their pioneering work, they investigated peripheral nerve stimulation (PNS), including the sciatic, ulnar, and occipital nerves, which exhibited promising results in reducing pain among patients [1]. This initial study laid the groundwork for subsequent research.

Building upon this foundation, Dr. Weiner and Dr. Reed, from the departments of Neurosurgery and Anesthesiology at Presbyterian Hospital of Dallas, Texas, conducted a seminal study in 1999. Focusing specifically on occipital neuralgia, they examined the application of ONS as a potential treatment. In their study, 13 patients underwent a total of 17 implant procedures. The implantation process involved the strategic placement of a subcutaneous electrode at the C1 level, spanning horizontally across the base of the occipital nerve trunk, guided by fluoroscopic imaging. Fig. 10.1 shows the positioning of their patient prior to generator placement. Notably, patients experienced both paresthesias and substantial pain relief, effectively targeting and alleviating the affected regions associated with occipital nerve pain [2].

ONS is thought to be effective through the inhibition of nociceptive activation in c-fibers and a-delta fibers. Therefore, the treatment has an antinocicep-

Interventional Management of Migraines and Other Headache Disorders. https://doi.org/10.1016/B978-0-443-23557-3.00010-3

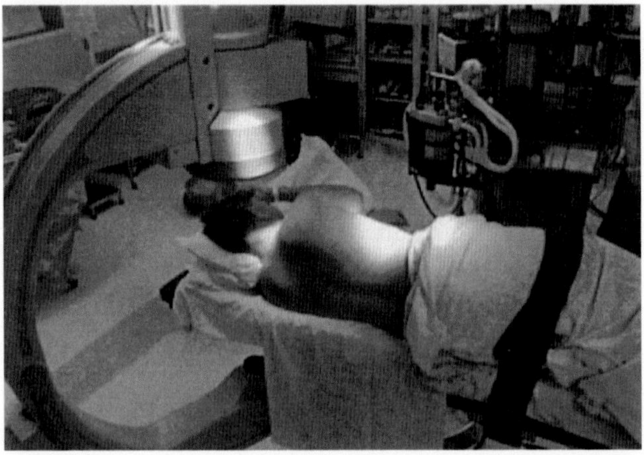

FIGURE 10.1 Lateral intraoperative positioning for generator receiver placement from [2]. From, R.L. Weiner, K.L. Reed, Peripheral neurostimulation for control of intractable occipital neuralgia, Neuromodulation 2 (3) (1999) 217–221. https://doi.org/10.1046/j.1525-1403.1999.00217.x.

tive effect in the region of the occipital nerves, along with trigeminal nerve innervation locations. There have also been interesting PET studies demonstrating a central mechanism of ONS where the treatment modulates the hypermetabolic brain areas involved in the pain generation [3]. This is further highlighted by studies showing significant blood flow increases on PET scans in the dorsal rostral pons, anterior cingulate cortex, and cuneus, which directly correlated to pain scores [4]. Subcutaneous peripheral neurostimulation of the occipital nerves is a viable option for refractory occipital neuralgia and other types of headache disorders.

Role of occipital nerve stimulation in headache treatment

Within the broader context of headache management, chronic headaches have been recognized to affect approximately 5% of the global population, causing significant disability [5]. Over the years, considerable progress has been made in the field of headache management. Indeed, a better understanding of the underlying mechanisms of specific headache types has led to the development of more targeted treatment approaches and improvements in existing technology. The utilization of biomarkers, such as genetic markers, inflammatory markers, and imaging markers, has played a pivotal role in identifying distinct headache subtypes, predicting treatment response, and facilitating the development of targeted therapies [6].

Pharmacological interventions, including monoclonal antibodies, and nonpharmacological approaches, such as cognitive-behavioral therapy, have expanded the range of treatment modalities available to patients with chronic headaches. These diverse options offer personalized and comprehensive strategies to address their unique needs and conditions. Chronic headaches management requires a multidisciplinary approach that includes a focus on the psychologic, social, and physical effects.

In this landscape of advancements, neuromodulation techniques, such as ONS, stand out as an additional treatment modality with considerable potential. The ability to target the occipital nerves directly provides a promising solution for individuals who have not responded to previous interventions or have encountered challenges with traditional pharmacological treatments. By offering a novel approach, ONS presents new possibilities for patients, promising relief from their chronic condition and improved quality of life.

INDICATIONS FOR OCCIPITAL NERVE STIMULATION

Types of headaches treated with occipital nerve stimulation

ONS has emerged as a valuable therapeutic option and has shown promise in various subtypes of headaches, mostly targeting headaches involved in the occipital region. ONS was initially employed in patients with refractory occipital neuralgia, which is a headache disorder characterized by radiating pain from the occiput. Occipital neuralgia has a frequency of 0.1%–4.7% in patients with headaches [7]. It is generally described as an intermittent, stabbing pain that lasts from seconds to minutes. The pain usually begins unilateral in the occipital region but often radiates throughout the head. There can also be tenderness to palpation in the area of the occipital nerve. The etiology is variable ranging from direct occipital nerve trauma, upper cervical root compression, rheumatologic, postsurgical, and idiopathic [8]. ONS has demonstrated safety and efficacy in treating this condition through multicenter studies [9].

Building on these findings, subsequent reports have investigated the application of ONS in other headache disorders, expanding its clinical utility. For instance, researchers and clinicians have explored the use of ONS for chronic migraines. Chronic migraine sufferers, who experience headaches for at least 15 days per month, have benefited from ONS as a potential adjunctive therapy to reduce the frequency and severity of their migraines [10]. Additionally, ONS has been investigated as a therapeutic option for cluster headaches, which are excruciating headaches that occur in clusters over a defined period, with studies indicating potential relief and improved quality of life in select patients treated with ONS [11]. In the realm of cervicogenic headaches, which stem from neck disorders or injuries, ONS has shown promise in relieving pain and improving overall headache outcomes. By targeting the occipital nerves, which play a role in cervicogenic headache pathophysiology, ONS offers a novel avenue for managing this condition [12].

Patient selection criteria

Patient selection criteria for ONS involve a thorough evaluation to identify individuals who are most likely to benefit from this treatment modality. Considering the invasiveness, risk of adverse events and cost associated with ONS, careful patient selection is crucial to maximize treatment efficacy and minimize potential risks. First, patients need a diagnosis of one of the headache disorders mentioned earlier in the chapter. Prior attempts at various noninvasive treatments, including medications, lifestyle modifications, physical therapy, and other interventions, should be documented to ensure that ONS is considered a next-step option after conventional treatments have proven ineffective. In most cases, patients undergo diagnostic nerve blocks using local anesthetics to temporarily block the occipital nerves. A positive response to these nerve blocks, where patients experience significant pain relief during the block, but the pain returns, can indicate a higher likelihood of benefiting from ONS. However, some studies have shown that an occipital nerve block (ONB) may not be a reliable predictor for the outcome of ONS. For further assurance, patients undergo a 1-week trial of stimulation prior to permanent implantation. A general consensus if the field with neuromodulation is that if the patient achieves a greater than 50% reduction in pain relief than a permanent stimulator is warranted. It is important to note that ONS is contraindicated in patients who fail the trial period. In addition, patients who are pregnant, have the presence of an Arnold–Chiari malformation, localized infection, or electronic implant are also contraindicated from receiving the device [13].

Advantages and disadvantages of occipital nerve stimulation

Advantages: First, ONS has shown promising results in reducing pain associated with chronic headaches demonstrating significant pain relief and improved quality of life in patients. Also, the treatment allows for a more focused and localized treatment approach, offering an alternative to pharmacological therapies, which may be ineffective or poorly tolerated by some patients. Indeed, ONS can be particularly beneficial for individuals who have not responded to or cannot tolerate medication due to side effects, drug interactions, or medical comorbidities. Additionally, ONS implantation is a reversible procedure, and the device can be adjusted or turned off if necessary.

Disadvantages: Prior to permanent implantation, patients typically undergo a trial period where temporary leads are placed to evaluate the effectiveness of ONS. While this allows for better patient selection, it adds an additional step to the treatment process. If the trial is successful, surgery is then required for permanent implantation of the device. With any surgery, there comes the risk of bleeding, infections, and adverse reactions to anesthesia that should be considered in patients who may be high-risk candidates. As a newer modality, the treatment can be costly, with many insurances requiring prior authorizations and documented failure of conventional therapies before approval is granted. Additionally, there are complications that can occur from the implanted device. Some of those complications include lead migration, device failure, or inadvertent shocks. Though these complications are rare, they are important to consider prior to making a decision to pursue ONS.

Techniques for occipital nerve stimulation

OVERVIEW OF TECHNIQUE

The technique of ONS was first described by Weiner and Reed in 1999, which outlined the percutaneous ONS technique [2]. In the years following, various alterations to the technique have been described and performed, such as using cylindrical percutaneous electrode leads and an open surgical approach with flat paddle-type electrodes [14,15]. There is no universally accepted approach to the ONS technique; there are differing opinions regarding the number of contacts required for adequate stimulation, the choice of electrodes, direction of electrode placement, entry point and anchoring location, generator location (midaxillary, gluteal, abdominal, infraclavicular, etc. and anchor types [16]. In this section, we summarize the technique illustrated by Waldman, Atlas of Interventional Pain Management [17].

The patient is positioned in a seated, lateral decubitus or prone position, depending on comfort and ability to maintain the position for the length of the procedure (30–45 min). The skin of the occipital and suboccipital region, and the electrode exit site, is prepared with aseptic technique. The first cervical vertebra is identified with anteroposterior and lateral views. The skin and subcutaneous tissue is anaesthetized with local anesthetic and a punctate midline incision is made. The Tuohy needle is then inserted subcutaneously toward the midline and placed at the level of the first cervical vertebrae (Figs. 10.2–10.4). The needle and electrode are then made wet with saline to avoid damage to the electrode insulation as the physician prepares to insert the electrode. The stimulating electrode is then advanced through the Tuohy needle medially to overlie the occipital nerves (Fig. 10.5) [17].

To prepare for tunneling, the incision is extended superiorly and inferiorly about 0.5 cm to dissect the tissue away from the needle. The tunneling tool can then be advanced with the electrode advancing into the incision. The needle is

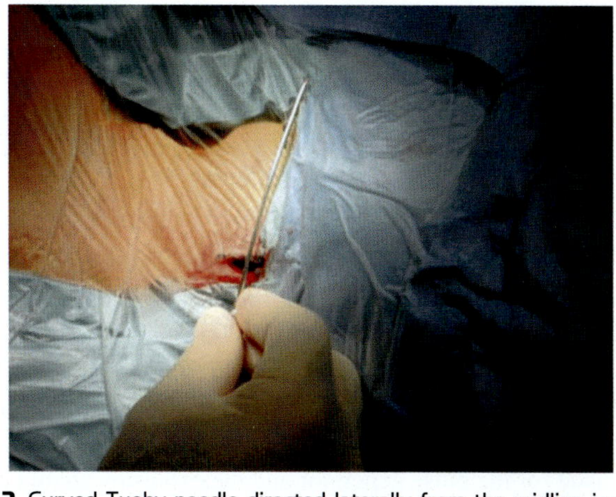

FIGURE 10.2 Curved Tuohy needle directed laterally from the midline incision. From S. Bruce, S.D. Waldman, Atlas of Interventional Pain Management, fifth ed., Anaesth. Intens. Care vol. 50 3 (2022). https://doi.org/10.1177/0310057x211038582.*

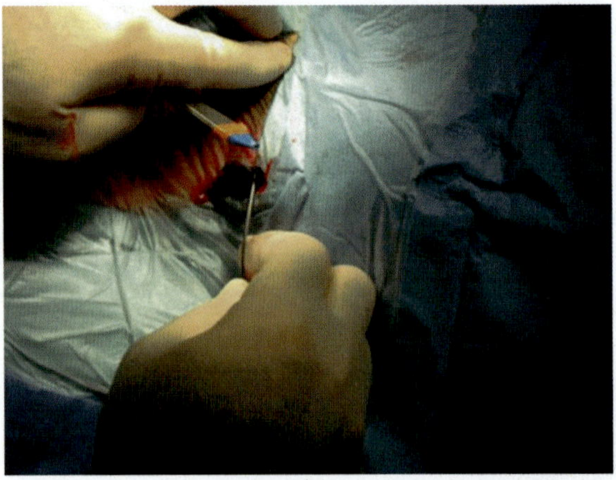

FIGURE 10.3 Subcutaneous needle insertion, carefully monitoring the needle tip position depth with the left index finger. From S. Bruce, S.D. Waldman, Atlas of Interventional Pain Management, fifth ed., Anaesth. Intens. Care vol. 50 3 (2022). https://doi.org/10.1177/0310057x211038582.*

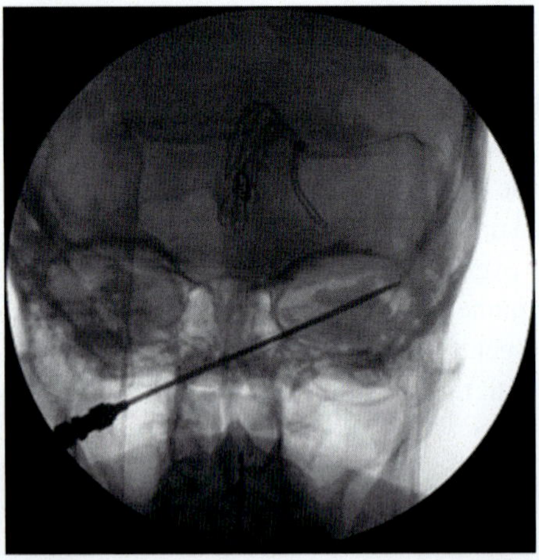

FIGURE 10.4 Fluoroscopic image demonstrating Tuohy needle overlying GON. From S. Bruce, S.D. Waldman, Atlas of Interventional Pain Management, fifth ed., Anaesth. Intens. Care vol. 50 3 (2022). https://doi.org/10.1177/0310057x211038582.*

then slowly withdrawn and removed, leaving the electrode in the tissue. The electrode is attached to the pulse generator through the sterile screening cable (Fig. 10.6). Trial stimulation is performed by asking the patient the type and location of stimulation and how the stimulation is affecting their pain. The ideal sensation is a stimulation pattern superimposed on the patient's baseline pain. If the position is adequate, the electrode is disconnected from the screening table. A pocket is made for the pulse generator and the extension should be connected to the electrode through the neck incision. The incisions are then closed and the

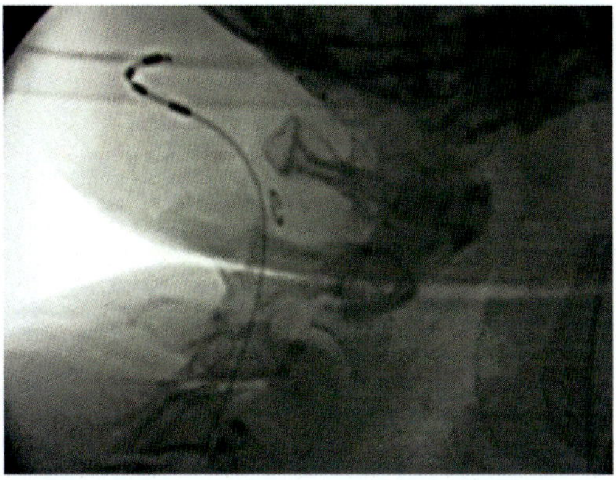

FIGURE 10.5 Fluoroscopic image confirming the correct electrode placement at C1 level. From S. Bruce, S.D. Waldman, Atlas of Interventional Pain Management, fifth ed., Anaesth. Intens. Care vol. 50 3 (2022). https://doi.org/10.1177/0310057x211038582.*

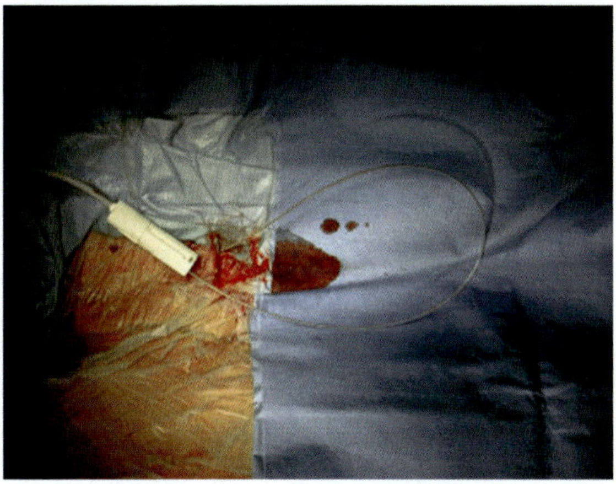

FIGURE 10.6 Electrode attached to pulse generator by the sterile screening cable. From S. Bruce, S.D. Waldman, Atlas of Interventional Pain Management, fifth ed., Anaesth. Intens. Care vol. 50 3 (2022). https://doi.org/10.1177/0310057x211038582.*

pulse generator is activated. Repeat images with posteroanterior and lateral fluoroscopic views are obtained to document correct position of the electrodes (Fig. 10.7). If lead movement is observed, a paddle-type electrode can then be implanted [17].

ANATOMIC LANDMARKS AND ULTRASOUND TECHNIQUES

The greater occipital nerve runs through the fascia below the superior nuchal ridge, running parallel with the occipital artery. The nerve supplies the posteromedial scalp and anteriorly to the vertex. The lesser occipital nerve runs cephalad and crosses the posterior border of the sternocleidomastoid muscle. It then divides into cutaneous branches that innervate the posterolateral scalp and the pinna of the ear (Fig. 10.8).

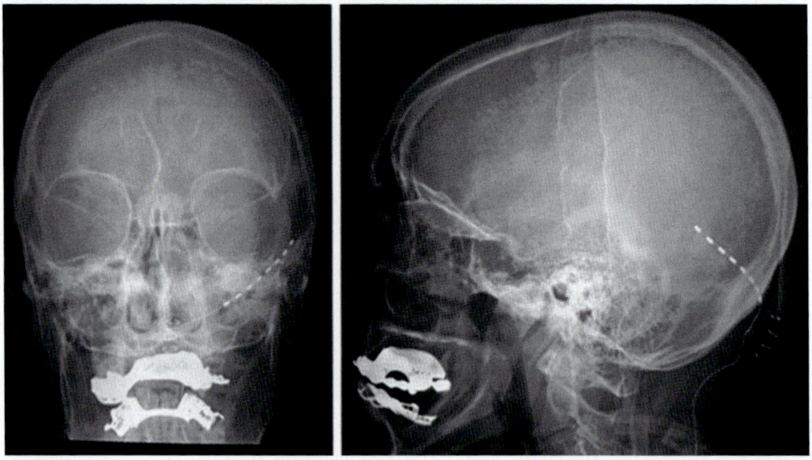

FIGURE 10.7 Anteroposterior (*left*) and lateral (*right*) radiographs after placement of occipital nerve stimulator with electrodes placed over occipital nerves. From S. Bruce, S.D. Waldman, Atlas of Interventional Pain Management, fifth ed., Anaesth. Intens. Care vol. 50 3 (2022). https://doi.org/10.1177/0310057x211038582.*

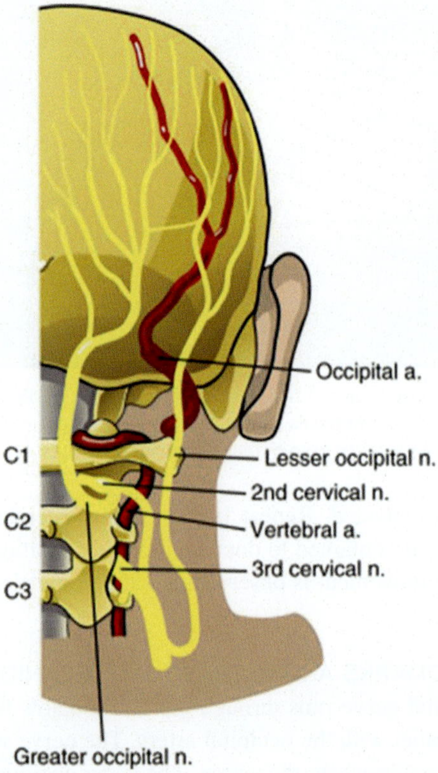

FIGURE 10.8 Anatomy of the occipital nerve. A = artery; N = nerve. Electrode insertion is generally performed under fluoroscopic guidance; however, ultrasound is a less commonly used alternative option that can be helpful in approximating electrode depth. From S. Bruce, S.D. Waldman, Atlas of Interventional Pain Management, fifth ed., Anaesth. Intens. Care vol. 50 3 (2022). https://doi.org/10.1177/0310057x211038582.*

EFFICACY AND SAFETY OF OCCIPITAL NERVE STIMULATION

Review of current evidence on the efficacy of occipital nerve stimulation for headache treatment

In the treatment of certain types of headaches, ONS has been found to be effective at reducing pain and improving quality of life. Occipital neuralgia, chronic migraines, cluster headaches, and cervicogenic headaches have all been successfully treated with ONS. The goals of ONS treatment in these headache disorders are to decrease the overall pain, frequency of attacks, number of headache days, and medication burden. A retrospective review conducted by Raoul et al. assessed the effectiveness of ONS to treat 60 patients with intractable occipital headaches. After 1 year of ONS treatment, the mean Visual Analogue Scale (VAS) had a statistically significant decrease of 72.2%, the mean Medication Quantification Scale (MQS) score decreased from 18 to 8.8 and there was a decrease of pain medication use by an average of 50% [8]. In a retrospective observational study with consecutive follow-up conducted by Diaz et al., the effect of ONS on 17 patients with drug-resistant chronic cluster headaches was evaluated. At the median follow-up time of 6 years there was a decrease of 30 to 22.5 weekly attacks in patients. Additionally, the VAS had a statistically significant decrease from 10 to seven at the end of follow up. Prophylactic oral medication use was reduced by 76.5% [18].

ONS has also been demonstrated to be effective in the treatment of chronic migraines. Dodick et al. performed a randomized, multicenter, double-blinded study looking at the long-term effects of ONS on chronic migraines over a 52-week period. 268 patients were enrolled and randomized to an active (with stimulation) or control group (no stimulation). Headache days were significantly reduced and almost 60% of patients achieved a 30% or greater reduction in headache days as measured by VAS [13]. Long term follow-up in studies is vital to assess the efficacy and inform decisions to implant ONS in a patient. Future studies with mean follow up times extending to 3 years and beyond are required to depict a comprehensive assessment of long-term patient outcomes. In addition to pain relief, reduced attack frequency, and medication burden, there can also be improvement in the functional and emotional burdens associated with chronic headaches.

Potential adverse effects and their management

Prior to ONS placement, it is vital to make sure there are no contraindications. Absolute contraindications to ONS placement include local infection and sepsis. Also, if a patient is on anticoagulant medication or has an underlying coagulopathy, this is an absolute contraindication because of the risk of bleeding [17].

Despite the benefits of ONS and other neuromodulation techniques, certain complications have been consistently reported across several studies. Lead or electrode displacement, infection, and hardware dysfunction are some of the most common reported complications. Raoul et al. reported that adverse events occurred in 12 out of the 60 patients that participated in their study (20%), with six patients presenting with electrode displacement or fracture (10%) and six patients presenting with cases of infection (10%) [8]. In another study involving ONS, 67 of the 105 patients experienced at least one complication with 29 patients requiring an additional surgery. The complications included infection (6%), lead migration (12%), fracture (4.5%), hardware dysfunction (8.2%), and local pain (20%) [19]. Dodick et al. reported a total of 183 device-procedure-related adverse events during their study, with 18 (8.6%) requiring hospitalization and 85 (40.7%) requiring surgical intervention. To limit lead migration causing further surgeries, additional contact points can be used. This allows for the leads to be reprogrammed after migration, instead of requiring a surgery to reinforce the lead placement [13]. Mechanical anchors and innovative procedures to relieve the strain on the leads can further prevent the complication of lead migration [20].

The risk of infection in ONS is overall rare. Patients who are immunocompromised are more at risk and ONS implantation should ideally be avoided. If infection does occur, the hardware is usually removed and drainage, culture, and antibiotic therapy are emergently started. The earlier the infection can be recognized, the better outcomes for the patient [17]. To prevent infection, the patient should be stratified according to risk, prophylactic antibiotics should be selected based on the respective hospital pathogens, using operating rooms with HEPA filters and laminar flow, decreasing the time of the surgical intervention, and using an occlusive dressing for 24–48 h [21]. Hardware failure is another potential complication that most commonly occurs from damage to the insulation of the stimulating electrode and loosening of the setscrews. These complications can be limited by wetting the electrode with sterile saline before advancing through the Tuohy needle and by carefully tightening the setscrews as the electrode is connected to the pulse generator or extension set [17]. While the surgical techniques of ONS have been improved, the complication rates are still relatively high and further technologic improvements are required.

Patient outcomes and satisfaction with the procedure

In the first large-scale, prospective controlled study evaluating ONS to treat chronic migraines over a 1-year period, there was a statistically significant reduction in headache days in patients treated with ONS. Importantly, more than two-thirds of patients treated with ONS reported satisfactory improvement in their headaches, and they had improved quality of life with overall satisfaction of the procedure. In patients that received ONS, over two thirds of patients reported that they would have the procedure again. The efficacy of ONS was sustained at 1 year follow up [13]. In a smaller retrospective open-label study that looked at 15 patients treated with ONS for medically refractory headaches, the results also reflected headache relief that was sustained on long term follow up. Specifically, headache frequency, disability (MIDAS), depression (Beck II), and headache severity demonstrated significant improvement at 19 month mean follow up. However, the positive results of this study need to be taken into context with the report of 60% of patients requiring surgical revision [22].

In another study that used bilateral ONS for 10 patients suffering from chronic cluster headaches, the frequency, duration and severity of cluster headaches was reduced by 90%. Importantly, 70% of the patients required less medication during the cluster headache attacks and all 10 of the patients reported overall improvement in their quality of life. Furthermore, there was evidence of improvement in the field of psychological comfort [23]. Overall, ONS has been proven to be effective in improving pain and quality of life in chronic headaches originating from the occipital region. However, the rate of complications for ONS remains high, despite improvements in techniques and equipment. Therefore, patients should be carefully selected for ONS and stratified based on risk of complications.

REFERENCES

[1] J.A. Picaza, S.E. Hunter, B.W. Cannon, Pain suppression by peripheral nerve stimulation: chronic effects of implanted devices, Stereotact. Funct. Neurosurg. 40 (2–4) (1977), https://doi.org/10.1159/000102446.

[2] R.L. Weiner, K.L. Reed, Peripheral neurostimulation for control of intractable occipital neuralgia, Neuromodulation 2 (3) (1999) 217–221, https://doi.org/10.1046/j.1525-1403.1999.00217.x.

[3] D. Magis, et al., Central modulation in cluster headache patients treated with occipital nerve stimulation: an FDG-PET study, BMC Neurol. 11 (2011), https://doi.org/10.1186/1471-2377-11-25.

[4] M.S. Matharu, T. Bartsch, N. Ward, R.S.J. Frackowiak, R. Weiner, P.J. Goadsby, Central neuromodulation in chronic migraine patients with suboccipital stimulators: a PET study, Brain 127 (1) (2004) 220–230, https://doi.org/10.1093/brain/awh022.

[5] P. Rizzoli, W.J. Mullally, Headache, Am. J. Med. 131 (1) (2018), https://doi.org/10.1016/j.amjmed.2017.09.005.

[6] H.G. Sutherland, C.L. Albury, L.R. Griffiths, Advances in genetics of migraine, J. Headache Pain 20 (1) (2019), https://doi.org/10.1186/s10194-019-1017-9.

[7] L. Kapural, N. Mekhail, S.M. Hayek, M. Stanton-Hicks, O. Malak, Occipital nerve electrical stimulation via the midline approach and subcutaneous surgical leads for treatment of severe occipital neuralgia: a pilot study, Anesth. Analg. 101 (1) (Jul. 2005) 171–174, https://doi.org/10.1213/01.ANE.0000156207.73396.8E.

[8] S. Raoul, et al., Efficacy of occipital nerve stimulation to treat refractory occipital headaches: a single-institution study of 60 patients, Neuromodulation 23 (6) (2020) 789–795, https://doi.org/10.1111/ner.13223.

[9] S. Palmisani, et al., A six year retrospective review of occipital nerve stimulation practice–controversies and challenges of an emerging technique for treating refractory headache syndromes, J. Headache Pain 14 (2013), https://doi.org/10.1186/1129-2377-14-67.

[10] P.J. Goadsby, J. Schoenen, M.D. Ferrari, S.D. Silberstein, D. Dodick, Towards a definition of intractable headache for use in clinical practice and trials, Cephalalgia 26 (9) (2006) 1168–1170, https://doi.org/10.1111/j.1468-2982.2006.01173.x.

[11] S. Miller, L. Watkins, M. Matharu, Treatment of intractable chronic cluster headache by occipital nerve stimulation: a cohort of 51 patients, Eur. J. Neurol. 24 (2) (2017), https://doi.org/10.1111/ene.13215.

[12] E.A. Melvin, F.R. Jordan, R.L. Weiner, D. Primm, Using peripheral stimulation to reduce the pain of C2-mediated occipital headaches: a preliminary report, Pain Physic. 10 (3) (2007) 453–460, https://doi.org/10.36076/ppj.2007/10/453.

[13] D.W. Dodick, et al., Safety and efficacy of peripheral nerve stimulation of the occipital nerves for the management of chronic migraine: long-term results from a randomized, multicenter, double-blinded, controlled study, Cephalalgia 35 (4) (2015) 344–358, https://doi.org/10.1177/0333102414543331.*.
[14] R.L. Jones, Occipital nerve stimulation using a medtronic resume II® electrode array, Pain Physic. 6 (4) (2003) 507–508, https://doi.org/10.36076/ppj.2003/6/507.
[15] M.Y. Oh, J. Ortega, J.B. Bellotte, D.M. Whiting, K. Aló, Peripheral nerve stimulation for the treatment of occipital neuralgia and transformed migraine using a C1-2-3 subcutaneous paddle style electrode: a technical report, Neuromodulation 7 (2) (2004) 103–112, https://doi.org/10.1111/j.1094-7159.2004.04014.x.
[16] K.V. Slavin, E.D. Isagulyan, C. Gomez, D. Yin, Occipital nerve stimulation, Neurosurg. Clin. N. Am. 30 (2) (Apr. 2019) 211–217, https://doi.org/10.1016/j.nec.2018.12.004.
[17] S. Bruce, S.D. Waldman, Atlas of Interventional Pain Management, fifth ed., Anaesth. Intens. Care vol. 50 3 (2022). https://doi.org/10.1177/0310057x211038582.*.
[18] J. Díaz-de-Terán, et al., Occipital nerve stimulation for pain modulation in drug-resistant chronic cluster headache, Brain Sci. 11 (2) (2021), https://doi.org/10.3390/brainsci11020236.
[19] A. Leplus, et al., Long-term efficacy of occipital nerve stimulation for medically intractable cluster headache, Neurosurgery 88 (2) (2021) 375–383, https://doi.org/10.1093/neuros/nyaa373.
[20] S. Falowski, D. Wang, A. Sabesan, A. Sharan, Occipital nerve stimulator systems: review of complications and surgical techniques, Neuromodulation 13 (2) (2010) 121–125, https://doi.org/10.1111/j.1525-1403.2009.00261.x.
[21] T.R. Deer, D.A. Provenzano, Recommendations for reducing infection in the practice of implanting spinal cord stimulation and intrathecal drug delivery devices: a physician's playbook, Pain Physic. 16 (3) (2013) E125–E128, https://doi.org/10.36076/ppj.2013/16/e125.
[22] T.J. Schwedt, D.W. Dodick, J. Hentz, T.L. Trentman, R.S. Zimmerman, Occipital nerve stimulation for chronic headache - long-term safety and efficacy, Cephalalgia 27 (2) (2007) 153–157, https://doi.org/10.1111/j.1468-2982.2007.01272.x.
[23] O.M. Mueller, C. Gaul, Z. Katsarava, H.C. Diener, U. Sure, T. Gasser, Occipital nerve stimulation for the treatment of chronic cluster headache lessons learned from 18 months experience, Zentralbl. Neurochir. 72 (2) (2011) 84–89, https://doi.org/10.1055/s-0030-1270476.

FURTHER READING

[1] I. Skaribas, K. Aló, Ultrasound imaging and occipital nerve stimulation, Neuromodulation 13 (2) (2010) 126–130, https://doi.org/10.1111/j.1525-1403.2009.00254.x.

Chapter | Eleven

Supraorbital nerve stimulation

Christopher L. Robinson[1], Cyrus Yazdi[2], Thomas T. Simopoulos[2], Alan David Kaye[3], Ivan Urits[4], Jamal J. Hasoon[5], Vwaire Orhurhu[6], Sait Ashina[2,7], Moises Dominguez[8,*]

[1]Department of Anesthesiology, Perioperative, and Pain Medicine, Harvard Medical School, Brigham and Women's Hospital, Boston, MA, United States; [2]Department of Anesthesiology, Critical Care, and Pain Medicine, Harvard Medical School, Beth Israel Deaconess Medical Center, Boston, MA, United States; [3]LSU Health Shreveport (Formerly LSU Health Sciences Center Shreveport), Department of Anesthesiology, Shreveport, LA, United States; [4]Southcoast Health, Brain and Spine, Wareham, MA, United States; [5]University of Texas Health Science Center, Department of Anesthesiology, Critical Care, and Pain Medicine, Houston, TX, United States; [6]University of Pittsburgh Medical Center, Susquehanna, Williamsport, PA, United States; [7]Department of Neurology, Beth Israel Deaconess Medical Center, Harvard Medical School, Boston, MA, United States; [8]Department of Neurology, Weill Cornell Medical College, New York Presbyterian Hospital, New York, NY, United States

INTRODUCTION

In 2014, the Food and Drug Administration (FDA) cleared the use of the first electrical transcutaneous stimulation device, Cefaly, for the prevention of episodic migraine by targeting the supraorbital nerve (Fig. 11.1) [1,2]. Amongst other such noninvasive devices including transcranial magnetic, vagal nerve, and combined occipital and trigeminal stimulation and remote electronic neuromodulation, Cefaly, also known as an external trigeminal nerve stimulator (eTNS), adds to the growing number of noninvasive migraine interventions that are able to provide patients with an low side effect option for migraine management [3–13]. The eTNS is placed in the middle of the forehead of the patient, and transcutaneous electrical stimulation is applied to the supraorbital and supratrochlear nerves, terminal branches of the trigeminal nerve ophthalmic division (Figs. 11.2 and 11.3) [2,14]. Given the limited side effects of this noninvasive device for episodic migraine prevention and acute treatment of migraine with and

*Contribution of Moises Dominguez will be in public domain.

Interventional Management of Migraines and Other Headache Disorders. https://doi.org/10.1016/B978-0-443-23557-3.00011-5

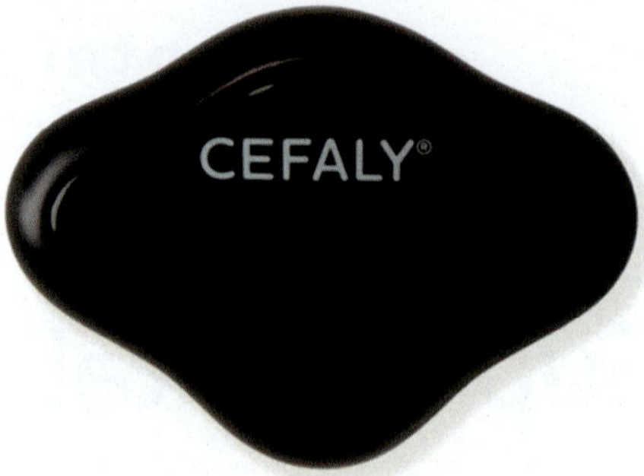

FIGURE 11.1 Cefaly: Image depicting the supraorbital electrical transcutaneous stimulation device. Courtesy of Cefaly.

without aura, the use of eTNS is overall a safe therapy for most patients but is contraindicated in patients with other implanted electrical devices such as pacemakers [2,14]. Overall, the device has reduced monthly migraine days up to 30% with a further reduction in acute migraine medications up to 37% for patients who have used Cefaly [2]. The eTNS offers migraine patients a safe modality for migraine management and often can be added as an adjunct potentially aiding in the decrease risk of medication overuse headache as its use has reduced migraine medication intake, and patients are willing to switch over to treatments with a lower side effect profile but similar efficacy [14,15].

MECHANISM OF ACTION

Despite numerous technologies utilizing electrical stimulation for the management of chronic pain, the mechanism of action remains to be fully elucidated.

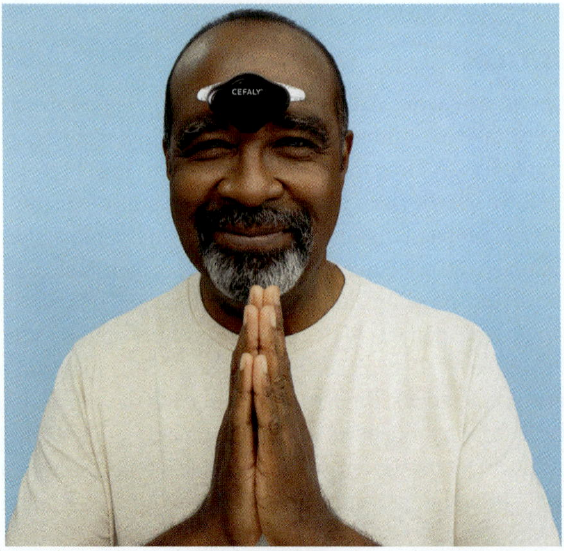

FIGURE 11.2 Cefaly on patient: Clinical image demonstrating the location of the transcutaneous supraorbital device, Cefaly, on a patient. Courtesy of Cefaly.

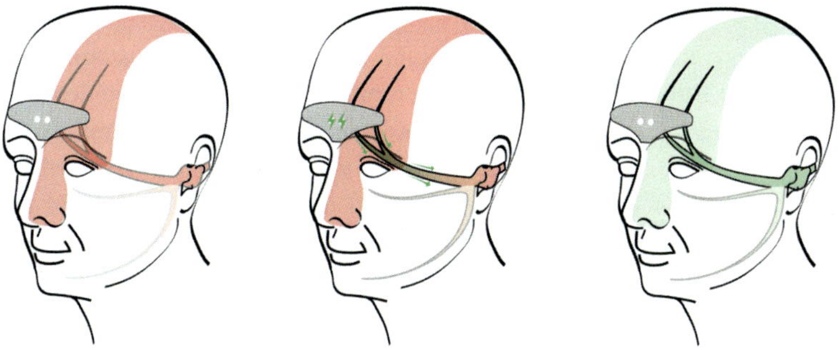

FIGURE 11.3 Supraorbital nerve stimulation with Cefaly: Illustration depicts the CNV_1. distribution of the trigeminal nerve (CNV) in *red* (pain) or *green* (pain relief). The supratrochlear nerve is located medial to the supraorbital nerve, which converge and form part of the frontal nerve a branch of CNV_1. When Cefaly is off, there is no simulation of the supraorbital nerve. Once activated, stimulation sends signals transcutaneously to the supraorbital nerve interfering with pain signaling.

In 1965, Ronald Melzack and Patrick Wall proposed the gate theory, which theorized that the regulation of pain with electrical stimulation results in the activation of Aβ-fibers [16–18]. These Aβ-fibers activate inhibitory interneurons that interfere with pain signaling by altering the transmission of pain in the ascending pathways [18]. Other mechanisms include the regulation of autonomic pathways by releasing neurotransmitters such as gamma-aminobutyric acid and adenosine, thereby enhancing central neural inhibition [17,19–27].

Furthermore, supraorbital stimulation may exert is role in what is believed to be the trigeminocervical complex (TCC) as evidence suggests the addition of supraorbital nerve stimulation to occipital nerve stimulation in patients with refractory chronic migraine may have a salutary effect [28]. Studies point toward the presence of a functional continuum between the trigeminal nucleus and cervical nerves, such as upper three cervical dorsal horns, form the TCC as the nociceptive afferents of both systems converge on the TCC [28–34]. The TCC can may also explain why only some fronto-temporally distributed pain will respond to occipital nerve stimulation [28,33].

ANATOMY

The supraorbital nerve, along with the supratrochlear nerve, is a branch of the ophthalmic division (cranial nerve V_1, CNV_1) of the trigeminal nerve (CNV). The supraorbital nerve can be located as it exits the orbit through the supraorbital foramen and can be palpated by finding the supraorbital notch [35]. The supraorbital nerve often has anatomical variations and commonly divided into two branches, superficial and deep, providing sensory innervation to the conjunctiva, periosteum, scalp to the vertex, forehead extending to the temporal and parietal areas, and upper eyelid [36]. The supratrochlear nerve is located medially to the supraorbital nerve and innervates the bridge of the nose, medial forehead, and medial aspect of upper eyelid [35].

PROCEDURE

The eTNS is a transcutaneous supraorbital nerve stimulation system, which is placed on the forehead to stimulate the supraorbital and supratrochlear nerves [1,37]. Specifically, the eTNS is positioned and applied over the forehead above the eyebrows (Fig. 11.2) [38]. The patient will first cleanse the forehead to remove any oils or debris and apply the self-adhesive electrode in the center of the forehead with the wings right above the eyebrows [38]. The device is then attached to the electrodes through magnetism, and the button is can be pressed to activate either acute migraine treatment or episodic migraine prophylaxis [38]. For acute migraine management, the button should be pressed once and the acute program will run for 60 minutes (maximum stimulation intensity: 16 milliamps; pulse width: 250 mseconds; pulse frequency: 100 Hz); for daily episodic migraine prophylaxis, press the button twice and the prevent program will run for 20 min (16 mA, 250 mseconds, 60 Hz) [38].

The FDA (cleared) indications for the eTNS include acute treatment of migraine with or without aura and the preventative treatment of episodic migraine in patients 18 years of age [38]. Contraindications include patients who have the presence of metallic or electronic devices in the head, pain of unknown origin, or implants such as cardiac pacemaker or implantable cardioverter-defibrillator, which can lead to interference and complications and/or death [38].

CLINICAL EFFICACY

Noninvasive

EPISODIC MIGRAINE PREVENTION

The first pilot study involving eTNS for the prevention of episodic migraine was performed on eight patients who had >1 year history of episodic migraine [39]. Patients were instructed to apply the eTNS for 20 minutes a day for 3 months [39]. Of the patients that continued after the trial period, four patients had a reduction in their migraine attack frequency and intensity with the mean migraine attack frequency showing a decrease from 3.9 to 2.8 attacks per month [39].

The first double-blind, randomized, sham-controlled trial to be performed using the eTNS was conducted on 67 patients with episodic migraine who had ^{3} two migraine attacks/month [37,40]. The patients were randomized 1:1 to either treatment arm and stimulated daily for 20 minutes for a total of 3 months [37]. In the treatment arm, the mean number of migraine days ($P = .023$), monthly migraine attacks ($P = .044$), and monthly acute migraine treatment ($P = .007$) were reduced by 29.5%, 18.7%, 37%, respectively [37]. Overall, the therapeutic gain in the supraorbital treatment arm was 26%, in line with other preventative medications and noninvasive device migraine management [37]. When 2313 eTNS users were surveyed, over 53% of users were satisfied with the use of it for the prevention of episodic migraine, consistent with the aforementioned study such that 70.6% of patients were satisfied [8,37,41,42]. Overall, side effects are minimal and limited to sedation, headache after session, forehead redness, nausea, and the very rare, one in 1000 allergy to the electrode adhesive gel. [1,38,43].

When eTNS was added to flunarizine for the prevention of episodic migraine for 3 months, the >50% responder rate was 78.43% compared with when either was used alone (46.15% and 39.22%, respectively) [44]. By adding eTNS to prophylactic episodic migraine medication, eTNS was able to increase the efficacy of prophylaxis while not increasing the side effects as there was no difference between the combined group and flunarizine alone [44]. When compared with prophylactic medication such as topiramate, topiramate did have a higher efficacy, but the side effect profile resulted in a quarter of the patients stopping the medication [45]. Whereas eTNS resulted in minimal side effects and was well tolerated [45].

ACUTE MIGRAINE TREATMENT

To assess the clinical efficacy and safety of eTNS for acute migraine management, an open label trial with an intention-to-treat analysis was performed on 30 patients who were experiencing an acute migraine attack with or without aura [46]. Patients were treated for 1 hour with eTNS resulting in a 57.1% ($P < .001$) and 52.4% ($P < .001$) reduction in mean pain intensity one and 2 hours after treatment, respectively [46]. During the 2 hours period after treatment, not a single patient required acute migraine medication, and no adverse events were recorded [46]. Additionally, eTNS demonstrated clinical efficacy in a randomized, double-blind, sham-controlled trial involving 538 patients with a diagnosis of migraine with or without aura who had two to eight migraine headache days and were randomized to either a 2-hours or sham stimulation at home [47]. Patients were given a unit and asked to treat me acute migraine in the 2-month period [47]. At 2 hours, the proportion of patients with pain freedom using eTNS was 7.2% higher than the sham with the sham having 18.3% with pain freedom ($P = .043$), 14.3% higher for pain relief at 2 hours with the sham being 55.2% ($P = .001$), and sustained pain freedom and relief at 24 hours was 7.0% and 11.5% higher than the sham with the sham being 15.8% ($P = .039$) and 34.4% ($P = .006$), respectively, with no adverse events in either arm [47].

In another at home study, the use of eTNS for the management of acute migraine was evaluated in a single-center, open-label study with 59 patients who self-administered a 2-hours stimulation session within 4 hours of onset of an acute moderate-to-severe in intensity migraine and did not use acute migraine medication for that episode [48]. Of the 59 patients, 48 patients were included in the analysis due to eligibility, and after a 2 hours stimulation session, 35.4% were pain free with 25.0% of patients maintaining pain freedom at 24 hours; only 15 patients reported adverse events with the most common yet minor and reversible being forehead paresthesia [48]. In a second multicenter RCT, 109 patients who were experiencing an active, acute migraine attack with or without aura were enrolled to evaluate the efficacy in acute migraine attacks [49]. After a 1-h stimulation, patients in the treatment arm had a reduction of 59% compared with 30% in the sham arm ($P < .0001$) [49]. Overall, the side effects from Cefaly are minor and include forehead paresthesia (2.03%), sedation (0.82%), headache (0.52%), and local skin allergy to the electrode gel (0.09%) [8].

*CHRONIC MIGRAINE PREVENTION (*INVESTIGATIONAL*)*

In a multicenter, open-label study, the clinical efficacy of eTNS was evaluated in patients with chronic migraine with and without medication overuse for the prevention of chronic migraine [50]. 23 patients were enrolled with 19 patients completing the study due to four dropping out (3 patients had worsening headache and neck tension and one patient had keratoconjunctivitis) due to intolerance [50]. Patients were stimulated for 20 minutes daily for 4 months resulting in 35% of patients having a 50% reduction in monthly migraine days and 50% reduction in acute migraine medication [50]. The data supporting the efficacy of eTNS for the prevention of chronic migraine continues to grow with a second study evaluating demonstrating similar results but with a slightly lower efficacy [51]. Finally, in a single-center, open-label study, 25 patients were enrolled with four not completing due to lack of perceived clinical efficacy (3 patients) and to lack of reliability (1 patient) [52]. The study included a per-protocol analysis (21 patients) and intention-to-treat (24 patients) analysis [52]. In the per-protocol analysis, there was a reduction in headache days of 1.20 days ($P = .05$), and in the intention-to-treat analysis, there was a reduction of 1.92 days ($P = .08$) [52].

Invasive (investigational)

Though supraorbital nerve blocks are used for migraine management, the use of invasive peripheral nerve stimulation (PNS) for migraine management remains to be fully explored and is currently considered investigational and off-label (Fig. 11.4) [53,54]. Despite the invasiveness, there remains a cohort of patients who remain refractory to other treatment options [55]. Since 1967, there have been reports of invasive PNS for treatment of facial pain with the first report being of a patient with trigeminal neuralgia in which the infraorbital nerve was stimulated for 5 minutes with a pair of wires inserted into the infraorbital foramen; fortunately, the stabbing facial pain could not be reproduced by lightly brushing the area for a total of 17 minutes in this patient [17,27]. In a case report being first of its kind, a 57-year-old man with posttraumatic trigeminal neuropathic pain and secondary dystonia following the orofacial injury was treated with invasive supraorbital stimulation [56]. Electrodes were implanted percutaneously at the supraorbital notch along the brow arch [56]. Upon stimulation, the patient had immediate resolution of his symptoms for 18 months [56].

Despite the clinical efficacy of invasive occipital nerve stimulation, there remains a subset of patients that do not respond to occipital nerve stimulation raising the questions: why do these patients not respond to occipital nerve stimulation or is it that the distribution of the migraine may not be in the distribution of the occipital nerve [33,57–66]. Given that occipital nerve stimulation was originally used to treat refractory occipital neuralgia, occipital nerve

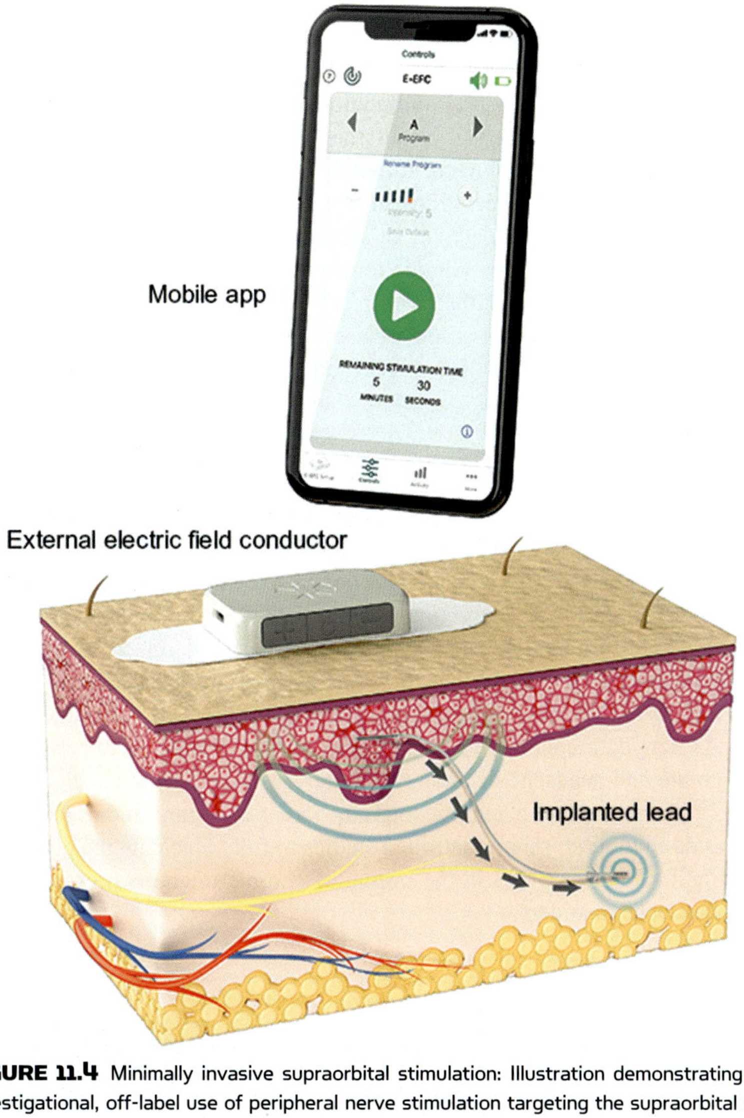

FIGURE 11.4 Minimally invasive supraorbital stimulation: Illustration demonstrating the investigational, off-label use of peripheral nerve stimulation targeting the supraorbital nerve. Courtesy of Bioaventus. Image not to be used for promotional purposes, only for educational and research purposes. For indications for use and other StimRouter relevant information, please visit www.stimrouter.com/safety-information/

stimulation development has led to its use for chronic migraine [67]. The studies on occipital nerve stimulation focus on three types of headaches: occipito-cervico, cluster, and chronic migraine with the average clinical efficacy being >85% for the first two and <50% for chronic migraine suggesting that occipital nerve stimulation may not be properly covering the entire source or there may be an additional site that may need to be targeted [28,33,68–78].

With this hypothesis in mind, combined invasive supraorbital and occipital nerve stimulation was performed on eight patients with chronic migraine, and the patients were treated with two programs – only occipital stimulation and combined occipital/supraorbital stimulation [28]. When stimulated with the combined setting, patients had a full therapeutic response with an inferior response with only occipital nerve stimulation [28]. As to whether the addition of the invasive supraorbital nerve stimulation enhances the mechanism of action through the common TCC pathway or supraorbital nerve stimulation covers a separate culprit causing the migraine remains to elucidated with the data suggesting the former [28]. Further research is necessary to fully elucidate the role of invasive supraorbital stimulation in chronic migraine prevention as it currently remains investigational and off-label.

CONCLUSION

eTNS, otherwise known as Cefaly, offers headache specialists another option in the prevention of episodic migraine and management of acute migraine with or without aura. Not only has eTNS demonstrated efficacy in populations in North America and Western Europe, it has also produced consistent results including a low side effect profile across various populations including Asia and Southern Europe demonstrating the wide applicability [7,79,80]. Furthermore, when eTNS was added on top of other therapies as an adjunct, its clinical efficacy only improved when compared with using it or the other therapies alone [81]. As new data continue to emerge, the role of eTNS in the management of migraine and headache disorders continues to expand with promising early results in the prevention of chronic migraine [50–52]. Future studies are needed such as double-blind, sham-controlled studies to fully elucidate the clinical efficacy and safety for eTNS in the setting of chronic migraine prevention and other headache disorders.

REFERENCES

[1] M.A.L. Johnson, D.E. Kuruvilla, External trigeminal nerve stimulation as a non-pharmacological option for the prevention and acute treatment of migraine, Neurology 18 (1) (2022) 22, https://doi.org/10.17925/usn.2022.18.1.22.

[2] D. Tepper, Transcutaneous supraorbital NeuroStimulation (tSNS), Headache J. Head Face Pain 54 (8) (2014) 1415–1416, https://doi.org/10.1111/head.12423.

[3] P. Barbanti, L. Grazzi, G. Egeo, A.M. Padovan, E. Liebler, G. Bussone, Non-invasive vagus nerve stimulation for acute treatment of high-frequency and chronic migraine: an open-label study, Italy J. Headache Pain 16 (1) (2015), https://doi.org/10.1186/s10194-015-0542-4.

[4] C.M. DeGiorgio, A. Shewmon, D. Murray, T. Whitehurst, Pilot study of Trigeminal Nerve Stimulation (TNS) for epilepsy: a proof-of-concept trial, Epilepsia 47 (7) (2006) 1213–1215, https://doi.org/10.1111/j.1528-1167.2006.00594.x.

[5] P.J. Goadsby, B.M. Grosberg, A. Mauskop, R. Cady, K.A. Simmons, Effect of noninvasive vagus nerve stimulation on acute migraine: an open-label pilot study, Cephalalgia 34 (12) (2014) 986–993, https://doi.org/10.1177/0333102414524494.

[6] R.B. Lipton, D.W. Dodick, S.D. Silberstein, J.R. Saper, S.K. Aurora, S.H. Pearlman, R.E. Fischell, P.L. Ruppel, P.J. Goadsby, Single-pulse transcranial magnetic stimulation for acute treatment of migraine with aura: a randomised, double-blind, parallel-group, sham-controlled trial, Lancet Neurol. 9 (4) (2010) 373–380, https://doi.org/10.1016/S1474-4422(10)70054-5.

[7] D. Magis, R. Jensen, J. Schoenen, Neurostimulation therapies for primary headache disorders: present and future, Curr. Opin. Neurol. 25 (3) (2012) 269–276, https://doi.org/10.1097/WCO.0b013e3283532023.

[8] D. Magis, S. Sava, T.S. d'Elia, R. Baschi, J. Schoenen, Safety and patients' satisfaction of transcutaneous supraorbital neurostimulation (tSNS) with the Cefaly® device in headache treatment: a survey of 2,313 headache sufferers in the general population, J. Headache Pain 14 (1) (2013), https://doi.org/10.1186/1129-2377-14-95.

[9] D. Magis, J. Schoenen, Advances and challenges in neurostimulation for headaches, Lancet Neurol. 11 (8) (2012) 708–719, https://doi.org/10.1016/s1474-4422(12)70139-4.

[10] Noninvasive Occipital and Trigeminal Neuromodulation Technology Cleared by FDA for Migraine Treatment. AHS.

[11] T.J. Schwedt, B. Vargas, Neurostimulation for treatment of migraine and cluster headache, Pain Med. (United States) 16 (9) (2015) 1827–1834, https://doi.org/10.1111/pme.12792.

[12] H. Yuan, T.Y. Chuang, Update of neuromodulation in chronic migraine, Curr. Pain Headache Rep. 25 (11) (2021), https://doi.org/10.1007/s11916-021-00988-7.

[13] S. Zhou, N. Hussain, A. Abd-Elsayed, R. Boulos, M. Hakim, M. Gupta, T. Weaver, Peripheral nerve stimulation for treatment of headaches: an evidence-based review, Biomedicines 9 (11) (2021) 1588, https://doi.org/10.3390/biomedicines9111588.

[14] F. Riederer, S. Penning, J. Schoenen, Transcutaneous supraorbital nerve stimulation (t-SNS) with the Cefaly® device for migraine prevention: a review of the available data, Pain Therap. 4 (2) (2015) 135–147, https://doi.org/10.1007/s40122-015-0039-5.

[15] R.M. Gallagher, R. Kunkel, Migraine medication attributes important for patient compliance: concerns about side effects may delay treatment, Headache 43 (1) (2003) 36–43, https://doi.org/10.1046/j.1526-4610.2003.03006.x.

[16] A. Abd-Elsayed, R.S. D'souza, Peripheral nerve stimulation: the evolution in pain medicine, Biomedicines 10 (1) (2022), https://doi.org/10.3390/biomedicines10010018.

[17] M.A.E. Elkholy, A. Abd-Elsayed, A.M. Raslan, Supraorbital nerve stimulation for facial pain, Curr. Pain Headache Rep. 27 (6) (2023) 157–163, https://doi.org/10.1007/s11916-023-01113-6.

[18] R. Melzack, P.D. Wall, Pain mechanisms: a new theory, Science 150 (3699) (1965) 971–979, https://doi.org/10.1126/science.150.3699.971.

[19] M. Aymanns, S.S. Yekta, J. Ellrich, Homotopic long-term depression of trigeminal pain and blink reflex within one side of the human face, Clin. Neurophysiol. 120 (12) (2009) 2093–2099, https://doi.org/10.1016/j.clinph.2009.08.027.

[20] H.A. Didier, P. Di Fiore, C. Marchetti, V. Tullo, F. Frediani, M. Arlotti, A.B. Giannì, G. Bussone, Electromyography data in chronic migraine patients by using neurostimulation with the Cefaly® device, Neurol. Sci. 36 (S1) (2015) 115–119, https://doi.org/10.1007/s10072-015-2154-9.

[21] External trigeminal nerve stimulation as a non-pharmacological option for the prevention and acute treatment of migraine - touchneurology.

[22] D. Magis, K. D'Ostilio, A. Thibaut, V. De Pasqua, P. Gerard, R. Hustinx, S. Laureys, J. Schoenen, Cerebral metabolism before and after external trigeminal nerve stimulation in episodic migraine, Cephalalgia 37 (9) (2017) 881–891, https://doi.org/10.1177/0333102416656118.

[23] M. Piquet, C. Balestra, S.L. Sava, J.E. Schoenen, Supraorbital transcutaneous neurostimulation has sedative effects in healthy subjects, BMC Neurol. 11 (2011), https://doi.org/10.1186/1471-2377-11-135Belgium.

[24] R. Radhakrishnan, E.W. King, J.K. Dickman, C.A. Herold, N.F. Johnston, M.L. Spurgin, K.A. Sluka, Spinal 5-HT2 and 5-HT3 receptors mediate low, but not high, frequency

TENS-induced antihyperalgesia in rats, Pain 105 (1) (2003) 205–213, https://doi.org/10.1016/s0304-3959(03)00207-0.

[25] J. Schoenen, Internal Medicine Review Migraine treatment with external trigeminal nerve stimulation migraine treatment with external trigeminal nerve stimulation: current knowledge on mechanisms, Int.Med. Rev. 3 (2017) 2017.

[26] E. Vecchio, E. Gentile, G. Franco, K. Ricci, M. de Tommaso, Effects of external trigeminal nerve stimulation (eTNS) on laser evoked cortical potentials (LEP): a pilot study in migraine patients and controls, Cephalalgia 38 (7) (2018) 1245–1256, https://doi.org/10.1177/0333102417728748.

[27] P.D. Wall, W.H. Swert, Temporary abolition of pain in man, Science 155 (3758) (1967) 108–109, https://doi.org/10.1126/science.155.3758.108.

[28] K.L. Reed, S.B. Black, C.J. Banta, K.R. Will, Combined occipital and supraorbital neurostimulation for the treatment of chronic migraine headaches: initial experience, Cephalalgia 30 (3) (2010) 260–271, https://doi.org/10.1111/j.1468-2982.2009.01996.x.

[29] T. Bartsch, P.J. Goadsby, Increased responses in trigeminocervical nociceptive neurons to cervical input after stimulation of the dura mater, Brain 126 (8) (2003) 1801–1813, https://doi.org/10.1093/brain/awg190.

[30] T. Bartsch, P.J. Goadsby, Stimulation of the greater occipital nerve induces increased central excitability of dural afferent input, Brain 125 (7) (2002) 1496–1509, https://doi.org/10.1093/brain/awf166.

[31] T. Bartsch, P.J. Goadsby, The trigeminocervical complex and migraine: current concepts and synthesis, Curr. Pain Headache Rep. 7 (5) (2003) 371–376, https://doi.org/10.1007/s11916-003-0036-y.

[32] R. De Agostino, B. Federspiel, E. Cesnulis, P.S. Sandor, High-cervical spinal cord stimulation for medically intractable chronic migraine, Neuromodulation 18 (4) (2015) 289–296, https://doi.org/10.1111/ner.12236Switzerland.

[33] C.A. Popeney, K.M. Aló, Peripheral neurostimulation for the treatment of chronic, disabling transformed migraine, Headache 43 (4) (2003) 369–375, https://doi.org/10.1046/j.1526-4610.2003.03072.x.

[34] A. Russo, A. Tessitore, F. Esposito, F. Di Nardo, M. Silvestro, F. Trojsi, R. De Micco, L. Marcuccio, J. Schoenen, G. Tedeschi, Functional changes of the perigenual part of the anterior cingulate cortex after external trigeminal neurostimulation in migraine patients, Front. Neurol. 8 (2017), https://doi.org/10.3389/fneur.2017.00282.

[35] N. Becser Andersen, G. Bovim, O. Sjaastad, The frontotemporal peripheral nerves. Topographic variations of the supraorbital, supratrochlear and auriculotemporal nerves and their possible clinical significance, Surg. Radiol. Anat. 23 (2) (2001) 97–104, https://doi.org/10.1007/s00276-001-0097-8.

[36] R. Haładaj, M. Polguj, M. Topol, Anatomical variations of the supraorbital and supratrochlear nerves: their intraorbital course and relation to the supraorbital margin, Med. Sci. Monit. 25 (2019) 5201–5210, https://doi.org/10.12659/MSM.915447.

[37] J. Schoenen, B. Vandersmissen, S. Jeangette, L. Herroelen, M. Vandenheede, P. Gerard, D. Magis, Migraine prevention with a supraorbital transcutaneous stimulator: a randomized controlled trial, Neurology 80 (8) (2013) 697–704, https://doi.org/10.1212/WNL.0b013e3182825055.

[38] CEFALY ® CONNECTED, A Medical Device for the Treatment of Migraine, 2022, p. 2022.

[39] P.Y. Gérardy, D. Fabry, A. Fumal, A pilot study on supra-orbital surface electrotherapy in migraine, Cephalalgia 29 (2009) 2009.

[40] I. Urits, R. Schwartz, D. Smoots, L. Koop, S. Veeravelli, V. Orhurhu, E.M. Cornett, L. Manchikanti, A.D. Kaye, F. Imani, G. Varrassi, O. Viswanath, Peripheral neuromodulation for the management of headache, Anesthesiol. Pain Med. 10 (6) (2020) 1–10, https://doi.org/10.5812/aapm.110515.

[41] S. Miller, L. Watkins, M. Matharu, Long-term outcomes of occipital nerve stimulation for chronic migraine: a cohort of 53 patients, J. Headache Pain 17 (1) (2016), https://doi.org/10.1186/s10194-016-0659-0.

[42] P. Verrills, C. Sinclair, A. Barnard, A review of spinal cord stimulation systems for chronic pain, J. Pain Res. 9 (2016) 481–492, https://doi.org/10.2147/JPR.S108884.
[43] V. Tiwari, S. Agrawal, Migraine and neuromodulation: a literature review, Cureus (2022), https://doi.org/10.7759/cureus.31223.
[44] L. Jiang, D.L. Yuan, M. Li, C. Liu, Q. Liu, Y. Zhang, G. Tan, Combination of flunarizine and transcutaneous supraorbital neurostimulation improves migraine prophylaxis, Acta Neurol. Scand. 139 (3) (2019) 276–283, https://doi.org/10.1111/ane.13050.
[45] G. Bussone, H.-C. Diener, J. Pfeil, S. Schwalen, Topiramate 100 mg/day in migraine prevention: a pooled analysis of double-blind randomised controlled trials, Int. J. Clin. Pract. 59 (8) (2005) 961–968, https://doi.org/10.1111/j.1368-5031.2005.00612.x.
[46] D.E. Chou, G.J. Gross, C.H. Casadei, M.S. Yugrakh, External trigeminal nerve stimulation for the acute treatment of migraine: open-label trial on safety and efficacy, Neuromodulation 20 (7) (2017) 678–683, https://doi.org/10.1111/ner.12623.
[47] D.E. Kuruvilla, J.I. Mann, S.J. Tepper, A.J. Starling, G. Panza, M.A.L. Johnson, Phase 3 randomized, double-blind, sham-controlled trial of e-TNS for the Acute treatment of Migraine (TEAM), Sci. Rep. 12 (1) (2022), https://doi.org/10.1038/s41598-022-09071-6.
[48] D. Kuruvilla, J.I. Mann, J. Schoenen, S. Penning, Acute treatment of migraine with external trigeminal nerve stimulation: a pilot trial, Cephalalgia Rep. 2 (2019) 251581631982990, https://doi.org/10.1177/2515816319829906.
[49] D.E. Chou, M. Shnayderman Yugrakh, D. Winegarner, V. Rowe, D. Kuruvilla, J. Schoenen, Acute migraine therapy with external trigeminal neurostimulation (ACME): a randomized controlled trial, Cephalalgia 39 (1) (2019) 3–14, https://doi.org/10.1177/0333102418811573.
[50] P. Di Fiore, G. Bussone, A. Galli, H. Didier, C. Peccarisi, D. D'Amico, F. Frediani, Transcutaneous supraorbital neurostimulation for the prevention of chronic migraine: a prospective, open-label preliminary trial, Neurol. Sci. 38 (S1) (2017) 201–206, https://doi.org/10.1007/s10072-017-2916-7.
[51] M. Birlea, S. Penning, K. Callahan, J. Schoenen, Efficacy and safety of external trigeminal neurostimulation in the prevention of chronic migraine: an open-label trial, Cephalalgia Rep. 2 (2019) 251581631985662, https://doi.org/10.1177/2515816319856625.
[52] C.M. Ordás, M.L. Cuadrado, J.A. Pareja, G. de-Las-Casas-Cámara, L. Gómez-Vicente, G. Torres-Gaona, B. Venegas-Pérez, B. Álvarez-Mariño, A. Diez Barrio, J. Pardo-Moreno, Transcutaneous supraorbital stimulation as a preventive treatment for chronic migraine: a prospective, open-label study, Pain Med. 21 (2) (2020) 415–422, https://doi.org/10.1093/pm/pnz119.
[53] C.A. Caputi, V. Firetto, Therapeutic blockade of greater occipital and supraorbital nerves in migraine patients, Headache 37 (3) (1997) 174–179, https://doi.org/10.1046/j.1526-4610.1997.3703174.x.
[54] S. Wahab, S. Kataria, P. Woolley, N. O'Hene, C. Odinkemere, R. Kim, I. Urits, A.D. Kaye, J. Hasoon, C. Yazdi, C.L. Robinson, Literature review: pericranial nerve blocks for chronic migraines, Health Psychol. Res. 11 (1) (2023), https://doi.org/10.52965/001c.74259.
[55] J.L. Natoli, A. Manack, B. Dean, Q. Butler, C.C. Turkel, L. Stovner, R.B. Lipton, Global prevalence of chronic migraine: a systematic review, Cephalalgia 30 (5) (2010) 599–609, https://doi.org/10.1111/j.1468-2982.2009.01941.x.
[56] J. Li, Y. Li, W. Shu, Case report: peripheral nerve stimulation relieves post-traumatic trigeminal neuropathic pain and secondary hemifacial dystonia, Front. Neurol. 14 (2023), https://doi.org/10.3389/fneur.2023.1107571.
[57] M. Anthony, Headache and the greater occipital nerve, Clin. Neurol. Neurosurg. 94 (4) (1992) 297–301, https://doi.org/10.1016/0303-8467(92)90177-5.
[58] Clinico Universitario Lozano Blesa. Occipital nerve stimulation for refractory chronic migraine: results of a long-term prospective study.
[59] N. Bogduk, Cervicogenic headache: anatomic basis and pathophysiologic mechanisms, Curr. Pain Headache Rep. 5 (4) (2001) 382–386, https://doi.org/10.1007/s11916-001-0029-7.

[60] N. Bogduk, The clinical anatomy of the cervical dorsal rami, Spine 7 (4) (1982) 319–330, https://doi.org/10.1097/00007632-198207000-00001.
[61] A.C. Brewer, T.L. Trentman, M.G. Ivancic, B.B. Vargas, A.M. Rebecca, R.S. Zimmerman, D.M. Rosenfeld, D.W. Dodick, Longa-term outcome in occipital nerve stimulation patients with medically intractable primary headache disorders, Neuromodulation 16 (6) (2013) 557–564, https://doi.org/10.1111/j.1525-1403.2012.00490.x.
[62] S. Miller, A.J. Sinclair, B. Davies, M. Matharu, Neurostimulation in the treatment of primary headaches, Pract. Neurol. 16 (5) (2016) 362–375, https://doi.org/10.1136/practneurol-2015-001298.
[63] P. Notaro, The effects of peripheral occipital nerve stimulation for the treatment of patients suffering from chronic migraine: a single center experience.
[64] K. Paemeleire, J.P. Van Buyten, M. Van Buynder, D. Alicino, G. Van Maele, I. Smet, P.J. Goadsby, Phenotype of patients responsive to occipital nerve stimulation for refractory head pain, Cephalalgia 30 (6) (2010) 662–673, https://doi.org/10.1111/j.1468-2982.2009.02022.x.
[65] S. Palmisani, A. Al-Kaisy, R. Arcioni, T. Smith, A. Negro, G. Lambru, V. Bandikatla, E. Carson, P. Martelletti, A six year retrospective review of occipital nerve stimulation practice–controversies and challenges of an emerging technique for treating refractory headache syndromes, J. Headache Pain 14 (2013) 67, https://doi.org/10.1186/1129-2377-14-67.
[66] J.R. Saper, D.W. Dodick, S.D. Silberstein, S. McCarville, M. Sun, P.J. Goadsby, Occipital nerve stimulation for the treatment of intractable chronic migraine headache: ONSTIM feasibility study, Cephalalgia 31 (3) (2011) 271–285, https://doi.org/10.1177/0333102410381142.
[67] R.L. Weiner, K.L. Reed, Peripheral neurostimulation for control of intractable occipital neuralgia, Neuromodulation 2 (3) (1999) 217–221, https://doi.org/10.1046/j.1525-1403.1999.00217.x.
[68] B. Burns, L. Watkins, P.J. Goadsby, Treatment of intractable chronic cluster headache by occipital nerve stimulation in 14 patients, Neurology 72 (4) (2009) 341–345, https://doi.org/10.1212/01.wnl.0000341279.17344.c9.
[69] B. Burns, L. Watkins, P.J. Goadsby, Treatment of medically intractable cluster headache by occipital nerve stimulation: long-term follow-up of eight patients, Lancet 369 (9567) (2007) 1099–1106, https://doi.org/10.1016/S0140-6736(07)60328-6.
[70] D. Magis, M. Allena, M. Bolla, V. De Pasqua, J.-M. Remacle, J. Schoenen, Occipital nerve stimulation for drug-resistant chronic cluster headache: a prospective pilot study, Lancet Neurol. 6 (4) (2007) 314–321, https://doi.org/10.1016/s1474-4422(07)70058-3.
[71] M.S. Matharu, T. Bartsch, N. Ward, R.S.J. Frackowiak, R. Weiner, P.J. Goadsby, Central neuromodulation in chronic migraine patients with suboccipital stimulators: a PET study, Brain 127 (1) (2004) 220–230, https://doi.org/10.1093/brain/awh022.
[72] M.Y. Oh, J. Ortega, J.B. Bellotte, D.M. Whiting, K. Aló, Peripheral nerve stimulation for the treatment of occipital neuralgia and transformed migraine using a C1-2-3 subcutaneous paddle style electrode: a technical report, Neuromodulation 7 (2) (2004) 103–112, https://doi.org/10.1111/j.1094-7159.2004.04014.x.
[73] Prospective case series using peripheral stimulation to reduce the pain of C2-mediated occipital headaches: a preliminary report, Pain Physician 10 (2007) 453–460.
[74] M.D. Rodrigo-Royo, J.M. Azcona, J. Quero, M. Cristina Lorente, P. Acín, J. Azcona, Peripheral neurostimulation in the management of cervicogenic headache: four case reports, Neuromodulation 8 (4) (2005) 241–248, https://doi.org/10.1111/j.1525-1403.2005.00032.x.
[75] T.J. Schwedt, D.W. Dodick, T.L. Trentman, R.S. Zimmerman, Occipital nerve stimulation for chronic cluster headache and hemicrania continua: pain relief and persistence of autonomic features, Cephalalgia 26 (8) (2006) 1025–1027, https://doi.org/10.1111/j.1468-2982.2006.01142.x.

[76] T.J. Schwedt, D.W. Dodick, J. Hentz, T.L. Trentman, R.S. Zimmerman, Occipital nerve stimulation for chronic headache - long-term safety and efficacy, Cephalalgia 27 (2) (2007) 153−157, https://doi.org/10.1111/j.1468-2982.2007.01272.x.

[77] K.V. Slavin, M.E. Colpan, N. Munawar, C. Wess, H. Nersesyan, Trigeminal and occipital peripheral nerve stimulation for craniofacial pain: a single-institution experience and review of the literature, Neurosurg. Focus 21 (6) (2006) E5, https://doi.org/10.3171/foc.2006.21.6.8.

[78] Trigeminal autonomic cephalalgias: current and future treatments, Headache J. Head Face Pain 47 (6) (2007) 981−986, https://doi.org/10.1111/j.1526-4610.2007.00840.x.

[79] D. Danno, M. Iigaya, N. Imai, H. Igarashi, T. Takeshima, The safety and preventive effects of a supraorbital transcutaneous stimulator in Japanese migraine patients, Sci. Rep. 9 (1) (2019), https://doi.org/10.1038/s41598-019-46044-8.

[80] A. Russo, A. Tessitore, F. Conte, L. Marcuccio, A. Giordano, G. Tedeschi, Transcutaneous supraorbital neurostimulation in "de novo" patients with migraine without aura: The first Italian experience, J. Headache Pain 16 (1) (2015), https://doi.org/10.1186/s10194-015-0551-3.

[81] P. Verrills, R. Rose, B. Mitchell, D. Vivian, A. Barnard, Peripheral nerve field stimulation for chronic headache: 60 cases and long-term follow-up, Neuromodulation 17 (1) (2014) 54−59, https://doi.org/10.1111/ner.12130.

Chapter | Twelve

Vagal nerve stimulation

Sohyun Kang[1], Roshan Santhosh[2], Shane Fuentes[3], Alan David Kaye[4,5]

[1]Department of Rehabilitation Medicine, Columbia University Medical Center, Weill Cornell Medical College, NewYork-Presbyterian Hospital, New York, NY, United States; [2]Department of Physical Medicine & Rehabilitation, Larkin Community Hospital, South Miami, FL, United States; [3]Department of Rehabilitation Medicine, Mayo Clinic School of Graduate Medical Education, Rochester, MN, United States; [4]Department of Anesthesiology, Louisiana State University Health Sciences Center, Shreveport, LA, United States; [5]LSU Health Shreveport (Formerly LSU Health Sciences Center Shreveport), Department of Anesthesiology, Shreveport, LA, United States

INTRODUCTION

Primary headaches can be disabling for patients, with migraines being the second leading cause of years lived with disability worldwide [1]. Related to debilitating pain experienced by many patients with primary headache disorders, finding effective treatments to alleviate their pain is paramount to ensuring patient success. Initially, headache pain relief from vagus nerve stimulation (VNS) was an incidental finding in patients being treated for intractable epilepsy with implantable VNS devices [2,3]. This prompted an investigation into VNS as a treatment for primary headache conditions. Findings from various experiments affirmed VNS, specifically noninvasive VNS (nVNS), as an effective treatment for some primary headache conditions, specifically cluster headaches and migraines [4−8].

Theories regarding the pathophysiology behind migraines and cluster headaches are incompletely understood. Migraines were initially believed to be caused primarily by cerebral and meningeal vasoconstriction. The current theories suggest involvement of the central and peripheral nervous symptoms through primary neuronal dysfunction related to genetic mutations of ion channels causing cortical hyperexcitability and depression, hypothalamic and brainstem activation, and the release of inflammatory peptides such as Calcitonin Gene-Related Peptide (CGRP) and pituitary adenylate cyclase-activating peptide [1,9]. Cluster headaches are thought to be caused by complex interactions between the trigeminovascular system, trigeminal-autonomic reflex, and posterior hypothalamus that respectively cause the unilateral, autonomic, and circadian/circannual symptoms that characterize the condition [10].

Interventional Management of Migraines and Other Headache Disorders. https://doi.org/10.1016/B978-0-443-23557-3.00012-7

In 2017, GammaCore, a handheld nVNS device that transcutaneously delivers electrical impulses to the vagus nerve, became the first FDA-cleared nVNS device for the acute treatment of cluster headaches, and was later expanded to include the prophylactic treatment of cluster headaches and the acute and prophylactic treatment of migraines [11]. While the exact mechanism by which nVNS treats these primary headaches is not fully understood, previous experiments suggest stimulation of the vagus nerve may alter pain perception by modulating neurotransmitter metabolism, particularly glutamate, and modulating cortical excitability [5]. Related to nVNS's favorable side effect profile and efficacy, nVNS shows promise for primary headache patients who are unable to tolerate side effects of pharmacotherapy and for patients whose headaches are refractory to conventional treatments [6]. In the present chapter, we review indications for VNS in primary headache disorders, patient selection, and VNS's advantages and disadvantages. We also discuss VNS techniques, as well as the efficacy and safety of the procedure.

INDICATIONS FOR VAGAL NERVE STIMULATION

Vagal nerve stimulation in the form of both surgically implanted cervical invasive vagal nerve stimulation and increasingly, noninvasive VNS (nVNS) have demonstrated clinical utility in management of primary refractory headache [12,13]. Types of headaches that have been identified as responders to treatment with VNS include chronic migraine (CM), high frequency episodic migraine (HFEM), chronic cluster headache (CCH), and episodic cluster headache [6,8,13–16]. Noninvasive VNS has been studied as an effective treatment modality for both acute migraine attacks as well as prophylactic treatment for migraine prevention [5,17]. Similarly, nVNS has also been investigated as potential prophylactic and abortive therapy for patients with chronic cluster headaches [14,15,18].

Patient selection criteria varied depending on the specific inclusion criteria of studies referenced, but generally recruitment was limited to patients with significant symptom intensity and frequency with established primary headache diagnoses who did not respond to standard of care treatment. For example, a landmark multicenter randomized controlled trial (RCT) examining nVNS for chronic migraine prophylaxis had inclusion criteria of: age 18–65 with previously diagnosed chronic migraine (CM) per International Classification of Headache Disorders criteria with >15 headache days/month for at least 3 months prior to enrollment in the trial [13]. Similarly, patients with history of aneurysm, intracranial hemorrhage, brain tumor/head trauma, previous surgery, abnormal anatomy at treatment site, or spine implants/hardware, known cardiovascular disease/abnormal ECGs were excluded [13]. Importantly, this study also controlled for confounding variables by excluding patients who had received onabotulinum toxin A injections in the previous 6 months as well as the use of any oral migraine prophylaxis medication in the month prior to enrollment in study [13]. Most RCTs reviewed in the scientific literature had similar inclusion criteria with selection of patients in defined age group with established clinical diagnosis of migraine or cluster headache with exclusion of those with advanced systemic disease such as cardiovascular disease, heart arrhythmias, systemic or neurological disorders, or those with implanted adjacent spine hardware or implanted electrical device [5,6].

Advantages of nVNS treatment included minimal adverse effects and significant reduction in headache days during a month as well as reduction in headache severity and frequency [13]. Noninvasive VNS was proven to be more cost effective and with improved quality of life compared with standard of care treatment for acute treatment of episodic migraine [19]. Disadvantages were limited in most trials with minimal nVNS-associated adverse effects in groups receiving nVNS [13].

TECHNIQUE

Noninvasive VNS is performed using a small, handheld device that applies electronic stimulation to the areas where the vagus nerve descends. The device can induce antinociception through transcutaneous vagus nerve stimulation (tVNS) at the external ear (auricular branch of vagus nerve) or the neck (the cervical branch of vagus nerve). There are several transcutaneous vagus nerve stimulator devices including NEMOS and gammaCore [12].

The NEMOS device is a noninvasive transauricular VNS (taVNS) that was approved in Europe for the treatment of pain in 2012. It is designed to provide stimulation at the concha of the outer ear, which receive sensory innervation from the auricular branch of the vagal nerve. It consists of a handheld, battery driven electrical stimulator connected to an ear electrode placed in contact with the skin of the concha. Impedance is measured automatically and insufficient electrode contact with the skin evokes an alarm. During stimulation, series of electrical pulses are applied to the skin of the concha [20]. Stimulus intensity can be individually fitted to elicit an appropriate response and can be adjusted as needed but it is usually recommended that stimulation lasts for 1–4 hours, 3–4 times a day [12].

The gammaCore device is a noninvasive transcervical VNS device (tcVNS) that stimulates the cervical branch of the vagus nerve. It was FDA cleared for the use in cluster headache as well as migraines. It has also received European clearance for the acute and prophylactic treatment of primary headaches and medication overuse headache. It stimulates the nerve with a 1 ms pulse repeated at 25 Hz. The gammaCore dosage for prevention is typically two stimulations (each 2 minutes in length) three times a day. For treatment, dosage is typically two stimulations immediately at headache onset, followed by two more at the 20-minute and 2-hour marks if necessary [21]. The pulse waveform was designed to penetrate the biological barrier (e.g., skin, tissue, nerve sheath) to stimulate the nerve. The stimulation intensity is self-controlled (up to 24 V and 60 mA) and lasts 2 minutes. It can be repeated multiple times a day without significant safety concerns [22].

As the electrodes' location and stimulation parameters differ in different studies, the various nVNS effects on pain may be a function of electrode position and stimulation parameters. Nonetheless, when appropriately stimulated, nVNS can interact with the vagus nerve system and elicit effective pain relief [23].

EFFICACY AND SAFETY OF VAGAL NERVE STIMULATION

The development of VNS, particularly nVNS for the management of primary headaches is a relatively novel indication in the field of neuromodulation, with FDA approval being granted for this indication in 2017–18 [6]. In this section we will review several excellent clinical trials, which demonstrate the clinical utility of nVNS in the management of several primary headache pathologies.

A multicenter, double blinded, sham controlled RCT for nVNS for patients with chronic migraine enrolled 59 participants who received low voltage (peak 24 V) stimulation of the unilateral cervical branch of the vagus nerve for two 2 minutes sessions at prefixed times, three times daily with the goal of acting as effective migraine prophylaxis. The results demonstrated statistically significant reduction in the average number of migraine days per month (Event study). Initial results from this trial were suggestive of a cumulative effect with increased duration of nVNS therapy, i.e., less headache days per month in each subsequent month spent enrolled in the trial. Study weaknesses and challenges to be addressed in future works include small sample sizes and high discontinuation rate. Given the nature of nVNS devices, there are challenges in designing an adequate sham device, which appears identical to real nVNS device while providing sensation of active treatment while avoiding production of treatment effects or adverse effects [13] (see Fig. 12.1).

A multicenter, randomized, double-blinded, sham controlled, parallel group study to investigate the efficacy of nVNS as acute abortive therapy for HFEM recruited 248 participants with established diagnosis of episodic migraine with or without aura. Patients were observed during three discrete 1 month long study stages of: observation of clinical baseline, double blinded period with the treatment group receiving bilateral 120 seconds stimulation of nVNS to right and left cervical branches of the vagal nerve within 20 minutes of migraine onset, as well as an open label period. The sham device produced a low frequency 0.1 Hz biphasic signal intended to be perceived by user without electrical stimulation of vagus nerve. Results were notable for statistically significant reduction in pain score with nVNS up to 2 h after migraine onset. There was also a statistically significant presence of pain free responders after nVNS at 30- and 60-minute post treatment, but not at 120 minutes. This trial had excellent statistical power and served as a foundational research trial demonstrating the clinical utility of nVNS as abortive therapy for HFEM in conjunction with conventional medical therapy with a high efficacy rate and minimal AE profile [6].

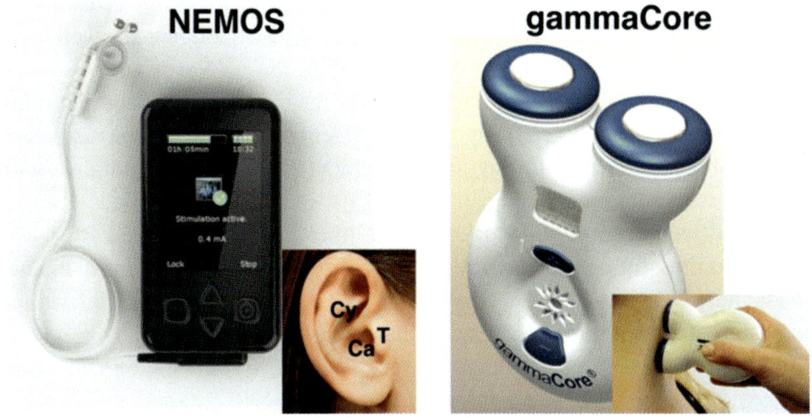

FIGURE 12.1 Two noninvasive VNS devices. NEMOS is a transauricular VNS device that stimulates the cymba chonca, which is innervated by the auricular branch of the vagus nerve. gammaCore is a transcervical, self-administered VNS device stimulating through the neck to the cervical vagus nerve [13].

A multicenter, randomized, double blinded, sham controlled study investigating the efficacy of nVNS for prophylactic treatment of HFEM recruited 477 patients to participate in a 4 week run in period, 12 weeks of double blind nVNS versus sham stimulation, and 24 week open label nVNS. Intervention consisted of nVNS of bilateral cervical branch vagal nerve versus sham treatment of low frequency 0.1 Hz biphasic current intended to be perceived without achieving vagus nerve stimulation. nVNS was provided 3 times a day at specified times, and rescue pharmacological acute migraine was permitted. Surprisingly, in this large, double blinded, randomized study with large sample size and strong statistical power, nVNS was not statistically superior to sham stimulation in reduction of migraine days/month. Further analysis revealed poor subject adherence, which may partly explain the poor performance of the nVNS group, although adjusted for >70% adherence, there is statistically significant reduction in migraine days in the group of patients who experience HFEM with aura. A subsequent study demonstrated significant parasympathetic activity from the signal generated by sham device, which was identified as a key limitation in the results of this study. These results demonstrate the study design challenges of creating a strong sham device that is believable to study participants without achieving vagal stimulation [8].

As follow up to the previous study, a multicenter, randomized, double blinded, sham controlled study investigating the efficacy of nVNS for prophylactic treatment of HFEM or chronic migraine with or without aura. This study enrolled 336 patients who were treated with nVNS as well as a true sham treatment (no electrical current) to cervical branch of vagal nerve on ipsilateral side of migraine symptoms, three times daily. The remainder of study design was identical to previous study, and results this time were notable for a reduction in number of migraine days/month, which was NOT significant as well as a decrease in abortive migraine medication use in the nVNS group. Although the results were not statistically significant for the primary endpoint, this study did demonstrate a statistically significant reduction in headache impact scores and significant reduction in migraine days/month in a subset of patients with migraine with aura [16].

A multicenter, randomized, double -blinded, sham controlled RCT was created to investigate the efficacy of nVNS for the acute treatment of cluster headaches. One hundred and fifty patients were enrolled in the trial, which consisted of a 1-month nVNS versus sham phase followed by a an open label 3 month nVNS period. Patient would receive either a controlled nVNS stimulus for three consecutive 2 minutes bursts to the R cervical vagal nerve branch at the onset of symptoms or sham treatment with of 0.1 mHz biphasic signal. The primary study end point was the proportion of patients who achieve a pain score of 0 or 1 after treatment as well as the duration of pain reduction up to an hour after the start of the cluster headache attack. Interestingly in this study, patients with intermittent cluster headache – defined as headaches lasting >7 days to 1 year with >1 month in between attacks responded more favorably to all measured outcome including response rate, sustained pain relief following nVNS, and percentage of patients who were responders compared with those with chronic cluster headaches lasting >1 year with less than 1 month in between relapses [14]. A similar study was repeated by the investigators, now separating the patient into chronic and episodic cluster headache subgroups, which confirmed the findings of the initial study, nVNS is an effective management tool for those with episodic cluster headaches but not for those with chronic cluster headaches compared with standard of care treatment. nVNS continues to have a favorable risk profile with minimal AEs and strong patient satisfaction across both studies [15].

An open label, single arm, multicenter study with patients experiencing HFEM or chronic migraine utilized nVNS as 120 seconds bursts at 3-minute intervals to the R cervical branch of the vagus nerve to achieve a 50% reduction in VAS (visual analog scale) score. An overwhelming majority of study participants 56.3% and 64.6% responded at the 1 and 2 h mark respectively [4]. A smaller portion of enrolled patients 17.6% and 22.9% reported no pain after nVNS at the 1- and 2-hour mark respectively. Patients who were noted to have HFEM experienced greater frequency of reduction in VAS after nVNS compared with the CM group. Study was limited by poor power secondary to small sample size of 50 patients [5].

The scientific literature currently does not suggest that there are significant adverse effects in clinical trials for various primary headache conditions with nVNS. Surgically implanted vagal nerve stimulators are associated with surgical implantation adverse effects including infection, bleeding, hardware malfunction and/or stimulation related adverse effects including chronic cough, voice disturbance, pain [24]. Mild adverse effects associated with nVNS include skin irritation at site of stimulation, stiff neck; however, multiple clinical trials with nVNS are suggestive of low rate of adverse effects, comparable with sham stimulation cohorts in trials [13–15,18].

Patient satisfaction in nVNS for chronic migraine was relatively high at 88.5%, with 46.7% of study participants achieving >50% treatment response as defined by reduction in average number of migraine days/month per study protocol [13]. Generally, patients reported >75% satisfaction with minimal mild adverse effects attributable to the use of nVNS.

There is a vast amount of data from well-designed clinical trials supporting the use of nVNS as an effective treatment modality for several etiologies of primary headache. Future studies to target different sites of vagus nerve stimulation (i.e., auricular), as well as presenting stronger evidence for the utility of nVNS for chronic cluster headaches are needed.

REFERENCES

[1] M.E. Lenaerts, K.J. Oommen, J.R. Couch, V. Skaggs, Can vagus nerve stimulation help migraine? Cephalalgia 28 (4) (2008) 392–395, https://doi.org/10.1111/j.1468-2982.2008.01538.x.

[2] R.M. Sadler, R.A. Purdy, S. Rahey, Vagal nerve stimulation aborts migraine in patient with intractable epilepsy, Cephalalgia 22 (6) (2002) 482–484, https://doi.org/10.1046/j.1468-2982.2002.00387.x.

[3] A.D. Nesbitt, J.C.A. Marin, E. Tomkins, M.H. Ruttledge, P.J. Goadsby, Non-invasive vagus nerve stimulation for the treatment of cluster headache: a case series, J. Headache Pain 14 (Suppl. 1) (2013) P231, https://doi.org/10.1186/1129-2377-1-S14-P231.

[4] P. Barbanti, L. Grazzi, G. Egeo, A.M. Padovan, E. Liebler, G. Bussone, Non-invasive vagus nerve stimulation for acute treatment of high-frequency and chronic migraine: an open-label study, J. Headache Pain 16 (1) (2015), https://doi.org/10.1186/s10194-015-0542-4.

[5] L. Grazzi, M. De Tommaso, G. Pierangeli, P. Martelletti, I. Rainero, S. Dorlas, et al., Noninvasive vagus nerve stimulation as acute therapy for migraine: the randomized PRESTO study, Neurology 91 (4) (2018) e364, https://doi.org/10.1212/WNL.0000000000005857.

[6] S.D. Silberstein, A.H. Calhoun, R.B. Lipton, B.M. Grosberg, R.K. Cady, S. Dorlas, K.A. Simmons, C. Mullin, E.J. Liebler, P.J. Goadsby, J.R. Saper, Chronic migraine headache prevention with noninvasive vagus nerve stimulation: the EVENT study, Neurology 87 (5) (2016) 529–538, https://doi.org/10.1212/WNL.0000000000002918.

[7] H.C. Diener, P.J. Goadsby, M. Ashina, M.A.M. Al-Karagholi, A. Sinclair, D. Mitsikostas, D. Magis, P. Pozo-Rosich, P. Irimia Sieira, M.J.A. Làinez, C. Gaul, N. Silver, J. Hoffmann, J. Marin, E. Liebler, M.D. Ferrari, Non-invasive vagus nerve stimulation (nVNS) for the preventive treatment of episodic migraine: the multicentre, double-blind, randomised, sham-controlled PREMIUM trial, Cephalalgia 39 (12) (2019) 1475–1487, https://doi.org/10.1177/0333102419876920.

[8] J. Yan, G. Dussor, Ion channels and migraine, Headache 54 (4) (2014) 619–639, https://doi.org/10.1111/head.12323.

[9] D.Y.T. Wei, J.J. Yuan Ong, P.J. Goadsby, Cluster headache: epidemiology, pathophysiology, clinical features, and diagnosis, Ann. Indian Acad. Neurol. 21 (5) (2018) S3, https://doi.org/10.4103/aian.AIAN_349_17.

[10] M. Mwamburi, E.J. Liebler, P.S. Staats, Patient experience with non-invasive vagus nerve stimulator: gammaCore patient registry, Am. J. Manag. Care 26 (1) (2020) S15, https://doi.org/10.37765/ajmc.2020.42545.
[11] J. Schoenen, A. Ambrosini, Update on noninvasive neuromodulation for migraine treatment—vagus nerve stimulation, Prog. Brain Res. 255 (2020) 249–274, https://doi.org/10.1016/bs.pbr.2020.06.009.
[12] M.S. Robbins, Diagnosis and management of headache: a review, JAMA, J. Am. Med. Assoc. 325 (18) (2021) 1874–1885, https://doi.org/10.1001/jama.2021.1640.
[13] H. Yuan, S.D. Silberstein, Vagus nerve and vagus nerve stimulation, a comprehensive review: Part II, Headache 56 (2) (2016) 259–266, https://doi.org/10.1111/head.12650.
[14] S.D. Silberstein, L.L. Mechtler, D.B. Kudrow, A.H. Calhoun, C. McClure, J.R. Saper, E.J. Liebler, E. Rubenstein Engel, S.J. Tepper, Non–invasive vagus nerve stimulation for the ACute treatment of cluster headache: findings from the randomized, double-blind, sham-controlled ACT1 study, Headache 56 (8) (2016) 1317–1332, https://doi.org/10.1111/head.12896.
[15] P.J. Goadsby, I.F. de Coo, N. Silver, A. Tyagi, F. Ahmed, C. Gaul, R.H. Jensen, H.C. Diener, K. Solbach, A. Straube, E. Liebler, J.C.A. Marin, M.D. Ferrari, Non-invasive vagus nerve stimulation for the acute treatment of episodic and chronic cluster headache: a randomized, double-blind, sham-controlled ACT2 study, Cephalalgia 38 (5) (2018) 959–969, https://doi.org/10.1177/0333102417744362.
[16] U. Najib, T. Smith, N. Hindiyeh, J. Saper, B. Nye, S. Ashina, C.K. McClure, M.J. Marmura, S. Chase, E. Liebler, R.B. Lipton, Non-invasive vagus nerve stimulation for prevention of migraine: the multicenter, randomized, double-blind, sham-controlled PREMIUM II trial, Cephalalgia 42 (7) (2022) 560–569, https://doi.org/10.1177/03331024211068813.
[17] F. Puledda, P.J. Goadsby, Current approaches to neuromodulation in primary headaches: focus on vagal nerve and sphenopalatine ganglion stimulation, Curr. Pain Headache Rep. 20 (7) (2016), https://doi.org/10.1007/s11916-016-0577-5.
[18] C. Gaul, H.C. Diener, N. Silver, D. Magis, U. Reuter, A. Andersson, E.J. Liebler, A. Straube, Non-invasive vagus nerve stimulation for PREVention and Acute treatment of chronic cluster headache (PREVA): a randomised controlled study, Cephalalgia 36 (6) (2016) 534–546, https://doi.org/10.1177/0333102415607070.
[19] M. Mwamburi, E.J. Tenaglia, P.S. Leibler, Staats, cost-effectiveness of noninvasive vagus nerve stimulation for acute treatment of episodic migraine and role in treatment sequence strategies, Am. J. Manag. Care 24 (2018).
[20] A. Straube, J. Ellrich, O. Eren, B. Blum, R. Ruscheweyh, Treatment of chronic migraine with transcutaneous stimulation of the auricular branch of the vagal nerve (auricular t-VNS): a randomized, monocentric clinical trial, J. Headache Pain 16 (1) (2015), https://doi.org/10.1186/s10194-015-0543-3.
[21] B. Blech, A.J. Starling, Noninvasive neuromodulation in migraine, Curr. Pain Headache Rep. 24 (12) (2020), https://doi.org/10.1007/s11916-020-00914-3.
[22] Y. Wang, G. Zhan, Z. Cai, B. Jiao, Y. Zhao, S. Li, A. Luo, Vagus nerve stimulation in brain diseases: therapeutic applications and biological mechanisms, Neurosci. Biobehav. Rev. 127 (2021) 37–53, https://doi.org/10.1016/j.neubiorev.2021.04.018.
[23] A.B. Antony, A.J. Mazzola, G.S. Dhaliwal, C.W. Hunter, Neurostimulation for the treatment of chronic head and facial pain: a literature review, Pain Physic. 22 (5) (2019) 447–477.
[24] I.S. Lendvai, A. Maier, D. Scheele, R. Hurlemann, T.M. Kinfe, Spotlight on cervical vagus nerve stimulation for the treatment of primary headache disorders: a review, J. Pain Res. 11 (2018) 1613–1625, https://doi.org/10.2147/JPR.S129202.

Chapter | Thirteen

Spinal cord stimulation for migraine headaches

Christopher L. Robinson[1,*], Cyrus Yazdi[2], Thomas T. Simopoulos[2], Jamal J. Hasoon[3], Sait Ashina[2,4], Vwaire Orhurhu[5,6], Alexandra Fonseca[1], Alan David Kaye[7,9], Moises Dominguez[8]

[1]Department of Anesthesiology, Perioperative, and Pain Medicine, Harvard Medical School, Brigham and Women's Hospital, Boston, MA, United States; [2]Department of Anesthesiology, Critical Care, and Pain Medicine, Harvard Medical School, Beth Israel Deaconess Medical Center, Boston, MA, United States; [3]University of Texas Health Science Center, Department of Anesthesiology, Critical Care, and Pain Medicine, Houston, TX, United States; [4]Comprehensive Headache Center, Department of Neurology, Beth Israel Deaconess Medical Center, Harvard Medical School, Boston, MA, United States; [5]University of Pittsburgh Medical Center, Susquehanna, Williamsport, PA, United States; [6]MVM Health, East Stroudsburg, PA, United States; [7]Department of Anesthesiology, Louisiana State University Health Sciences Center, Shreveport, LA, United States; [8]Department of Neurology, Weill Cornell Medical College, New York Presbyterian Hospital, New York, NY, United States; [9]LSU Health Shreveport (Formerly LSU Health Sciences Center Shreveport), Department of Anesthesiology, Shreveport, LA, United States

INTRODUCTION

Spinal cord stimulators (SCS) have undergone significant advancements in the goal of targeting persistent, intractable chronic pain; initially termed dorsal column stimulators, SCS was first used in 1967 followed by approval by the Federal Drug Administration (FDA) in 1989 [1−3]. Currently, SCS is approved by the FDA for the management of failed back surgery syndrome, complex regional pain syndrome, diabetic neuropathy, and nonsurgical back pain [1,4−6]. Despite approval for only four indications, research continues to explore the potential opportunities for which SCS can be potentially applied. Though currently investigational and off-label, an increasing number of studies have continued to support the ever-expanding role of SCS to manage chronic pain such as in the management of refractory chronic migraine (RCM) [7−14].

*Contribution of Moises Dominguez will be in public domain.

Interventional Management of Migraines and Other Headache Disorders. https://doi.org/10.1016/B978-0-443-23557-3.00013-9

The application of SCS employs the insertion of one to two electrodes into the epidural space in which electrical currents is generated to induce paresthesia, which serve as a substitute for pain [5,15]. Initially, patients undergo a trial of neuromodulation where temporary SCS leads are inserted into the epidural space and powered externally by an external pulse generator [5,15]. This trial assesses whether the patient achieves pain relief. If the trial proves successful, a more permanent subcutaneous internal pulse generator (IPG) is implanted along with one to two leads targeting the similar area as in the trial to provide long-term therapeutic benefits (Fig. 13.1).

Recent advances in SCS have led to the development of dorsal root ganglion (DRG) SCS, burst SCS, and high frequency 10 SCS (HF10) [5,16,17]. Despite the efficacy of SCS in alleviating intractable chronic pain, patients do report the unpleasant sensation of paresthesia; HF10 addresses the issue of unpleasant paresthesia by utilizing a waveform of 10,000 Hz at a subthreshold level [17]. Burst SCS applies conventional SCS parameters and delivers bursts of five pulses with reduced amplitude aiming to provide subthreshold stimulation while minimizing the occurrence of paresthesia yet still effectively alleviating pain [18,19].

In this chapter, therefore, we investigate recent findings for the investigational, off-label use of SCS for the management of RCM describing mechanisms, efficacy, and safety.

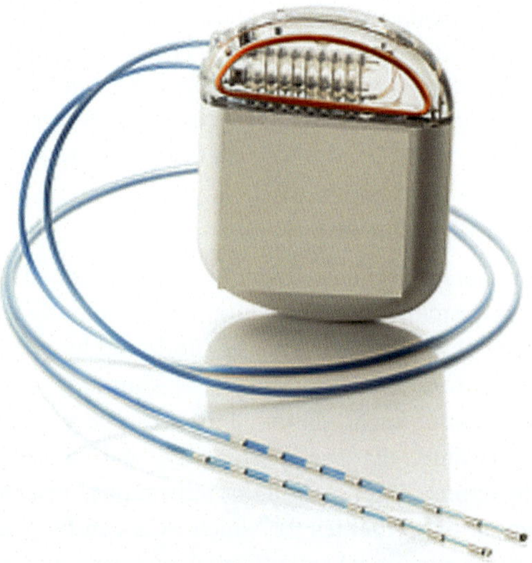

FIGURE 13.1 Spinal cord stimulator. SCS with two leads that are inserted in the epidural space along with the implantable pulse generator placed in the gluteal or paraumbilical region. Courtesy of 2023 Nevro Corp. SCS for migraine is investigational and off-label; image and content is used for educational purposes and not for promotional purposes.

ANATOMY OF THE SPINAL CORD

The spinal cord is protected by the spine. It consists of vertebral bodies interspaced with intervertebral discs, providing cushion and support with movement and prevents fracturing of the vertebrae. Each vertebra is formed by a vertebral body with bilateral pedicles that extend out of the body and connect the lamina, which join to form the spinous process (Fig. 13.2). The vertebral body, pedicles, and lamina form the spinal canal. Additionally, the pedicles form the neural foramen as it creates the lateral, superior, and inferior borders. Within the spinal canal, the first space encountered in which the SCS leads will be placed is the epidural space, which contains epidural fat and a venous plexus. Deep to the epidural space is the dura mater, which encases the spinal cord and is where cerebrospinal fluid is contained and flows cushioning the spinal cord from movement and preventing impact with the vertebrae [20,21].

The spinal cord is separated into equal parts by the dorsal median sulcus anteriorly and the ventral median fissure posteriorly with the central canal located as the name implies, centrally, serving the purpose to maintain continuity with the cerebral ventricular system in the cranial direction (Fig. 13.3). Surrounding the central canal is the gray matter, composed of nerve cell bodies, dendrites, and synaptic connections. Encircling the gray matter is the outer region referred to as white matter, consisting of ascending (dorsal column system and spinothalamic tracts) and descending nerve fibers (corticospinal, tectospinal, rubrospinal, vestibulospinal, and reticulospinal tracts), which form tracts that each serve their own purpose (Fig. 13.4).

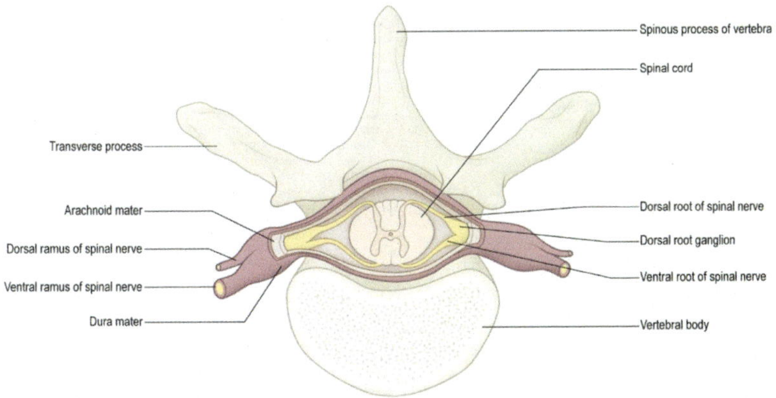

FIGURE 13.2 Vertebra features. Vertebral body connects to the lamina through the pedicle. The bilateral lamina merge to form the spinous process with the transverse process located lateral to the spinous process and projecting outward. The vertebra encases the spinal cord forming the spinal canal. Illustration taken from Fig. 8.1 in A.R. Crossman, et al. Neuroanatomy: an Illustrated Color Text. Please adapt the above image from an Elsevier publication into one for our publication. Since the SCS is placed usually in the thoracic region, I figured this would be better than the cervical region since then I would have to describe the different ones including C1 and C2. All figure descriptions are original.

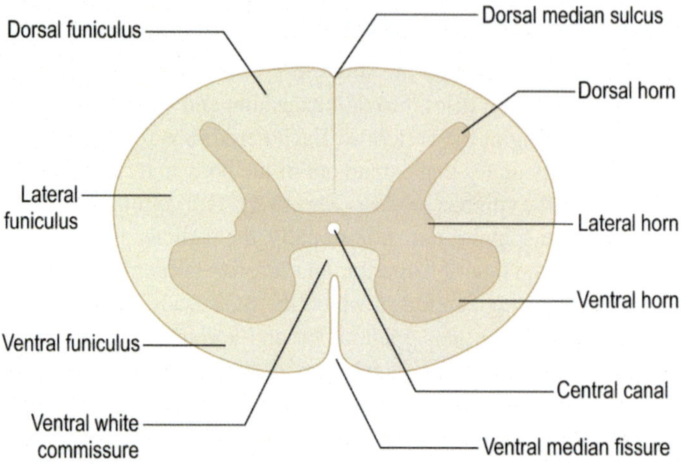

FIGURE 13.3 Spinal cord anatomy: The spinal cord is divided equally by the dorsal median sulcus anteriorly and the ventral median fissure posteriorly with the central canal located centrally maintaining continuity with the cerebral ventricular system in the cranial direction. Illustration taken from Fig. 8.8 in A.R. Crossman, et al. Neuroanatomy: an Illustrated Color Text. Please adapt the above image from an Elsevier publication into one for our publication.

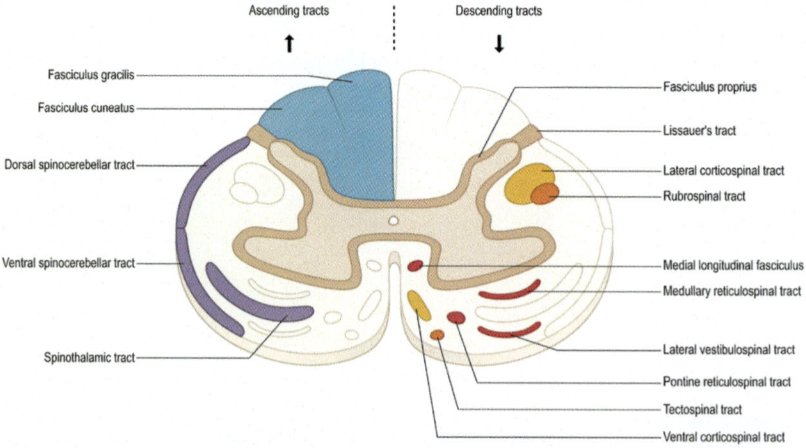

FIGURE 13.4 The dorsal column is arranged in a layered manner, representing the structures from caudal to rostral regions established in a medial to lateral fashion within the dorsal column. Illustration taken from Fig. 8.15 in A.R. Crossman, et al. Neuroanatomy: an Illustrated Color Text. Please adapt the above image from an Elsevier publication into one for our publication.

There exist nerve fibers that transverse the spinal cord itself called interneurons, which serve a function at only the spinal level such as spinal reflexes and pain modulation [22]. The transmission of signals from large afferent fibers such as Ab fibers activate these dorsal horn interneurons. These then in turn synapse on the presynaptic terminal of nociceptive ascending primary afferent fibers inhibiting the transmission of pain signaling superiorly [20,21].

Of the spinal cord tracts, the dorsal column remains the focus for SCS. The dorsal column is composed of the fasciculi cuneatus and gracilis, which are arranged in a layered manner, representing the structures covered from caudal to rostral regions established in a medial to lateral fashion within the dorsal column on each side (Fig. 13.4). The dorsal column on each side is located between the dorsal horn and median sulcus carrying signals such as proprioception and fine touch from the ipsilateral periphery, which enters the spinal cord through the dorsal roots. These first-order neurons continue to ascend until they synapse with second-order neurons at the level of the medulla oblongata (Fig. 13.5). The second-order neurons then decussate at the medulla and ascend superiorly as the medial lemniscus. Upon termination at the ventral posterolateral nucleus of the thalamus, the second-order neurons synapse on

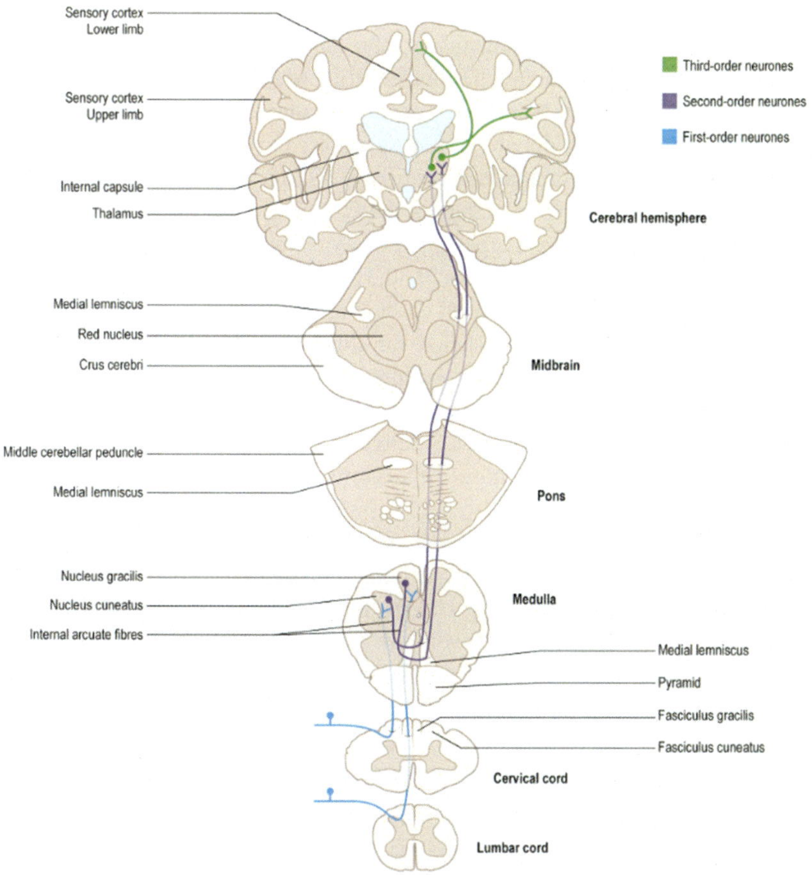

FIGURE 13.5 The pathway of the transmission through the ascending dorsal column to the somatosensory cortex located in the postcentral gyrus of the parietal lobe. Illustration taken from Fig. 8.16 in A.R. Crossman, et al. Neuroanatomy: an Illustrated Color Text. Please adapt the above image from an Elsevier publication into one for our publication.

third-order thalamocortical neurons, which project and terminate in the post-central gyrus of the parietal lobe, also called the somatosensory cortex [20,21].

MECHANISM OF ACTION

With the inception of the gate control theory by Ronald Melzack and Patrick Wall in 1965, SCS was introduced with the understanding that regulation of pain could occur at the level of the dorsal horn through Aβ-fibers, which serve to activate inhibitory interneurons that interfere with pain signaling [22]. Though having been first used nearly 60 years ago, the exact mechanism for how SCS provides pain relief remains to be fully elucidated but includes mechanisms as the aforementioned inhibitory effect on dorsal horn hyperexcitability due to release of either substance P, serotonin, or gamma-aminobutyric acid, activation of the dorsal horn cholinergic system and/or supraspinal descending pain modulating centers, inhibition of efferent sympathetic fibers, and suppression of the efferent sympathetic response [1,23–29].

Despite the proposed mechanisms, there may exist an additional separate mechanism of action for pain relief in the setting of migraine. Preclinical investigations indicate the presence of a functional continuum between the caudal trigeminal nucleus with upper cervical segments forming a trigeminocervical complex, implicated in cranial nociception [9,30,31]. Furthermore, stimulating the greater occipital nerve (a branch of the C2 spinal root) exerts a facilitatory effect on dural stimulations suggesting the involvement of a central mechanism at the level of second-order neurons [9,30,31]. When suboccipital steroid injections are performed for treatment of chronic migraine, sufferers experience relief further supporting the evidence for this network [32]. When directly stimulating the dorsal columns in the high-cervical region (C2-3) at the level of the medulla oblongata, analgesia efficacy may be just as efficacious as occipital nerve stimulation (ONS) as SCS delivers its effects in closer proximity to the trigeminal nucleus activating the occipital nerve through stimulation of the C2 nerve root [9].

Data suggest that when using paresthesia free modes, activation of only inhibitory neurons occurs in the dorsal horn without activation of excitatory interneurons, especially when using 10 kHz, further adding evidence to the inhibitory effect of SCS at the spinal level [33]. Furthermore, the lack of paresthesia provides patients the option to continue stimulation while asleep improving overall health status given the link between sleep quality and migraine [34–38].

CLINICAL EFFICACY FOR SCS (INVESTIGATIONAL)

Prior to the investigational use of SCS for RCM, ONS was performed when patients failed management for RCM as ONS has shown clinical efficacy, but it is not without issue, as it can have a higher complication rate than SCS [39–46]. Thus, the investigational use of SCS for RCM may not displace ONS but only adds to the ever growing arsenal of procedures that headache and chronic pain specialist have at their disposal. As the following studies reveal, SCS demonstrates promising results for pain alleviation in RCM patients.

When used for RCM and depending on the access to the epidural space, one to two SCS leads are inserted into the epidural space in the superior thoracic region and advanced superiorly to the high-cervical region (C2-3), and the IPG is implanted either in the paraumbilical or gluteal region (Fig. 13.6) [13]. The first reported use of SCS for headache disorders was the use in a patient suffering from chronic cluster headache (CCH) refractory to conservative management and minimally invasive treatment options such as greater occipital nerve (GON) and C2 nerve blocks, botulinum toxin injections, neurolysis of GON and C2 nerve roots, cryoablation, and cervical spondylodesis with bone interponate [8]. For the trial and permanent procedures, the SCS leads were inserted at the T2-3 epidural space and advanced to the inferior portion of the occipital bone [8]. The patient achieved sustained pain relief immediately after SCS placement and continued to prophylactically stimulate, providing sustained relief up to 24 h after stimulation; furthermore, the attack rate decreased from 8/day to 1/day every other day [8].

Given the success of the first patient, an additional seven patients with intractable CCH refractory to medical management were implanted with a high-cervical SCS and assessed after the an average trial period of 10 days [7]. All seven patients were then implanted with the permanent SCS and

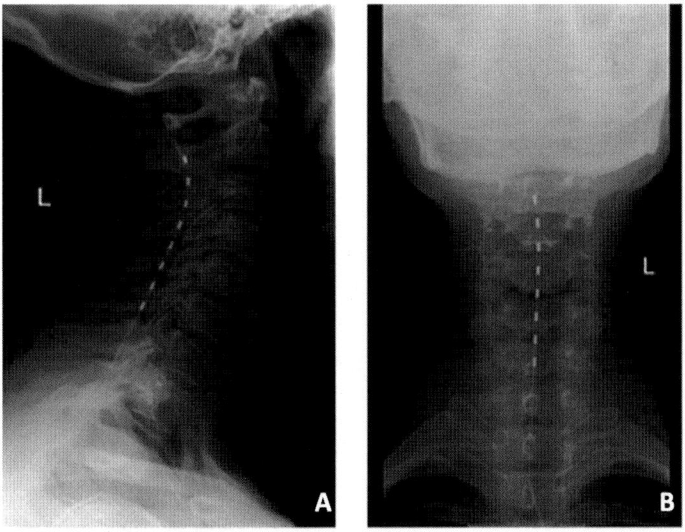

FIGURE 13.6 Fluoroscopic images of a single lead high-cervical SCS. (A) Anteroposterior and (B) lateral views of a single SCS lead placed in the cervical region for the investigational, off-label use for RCM. From A. Al-Kaisy, S. Palmisani, R. Carganillo, S. Wesley, D. Pang, A. Rotte, A. Santos, G. Lambru, Safety and efficacy of 10 kHz spinal cord stimulation for the treatment of refractory chronic migraine: a prospective long-term open-label study, Neuromodulation 25 (1) (2022) 103–113. doi: 10.1111/ner.13465

prospectively followed for an average of 23 months [7]. All patients experienced a near instant improvement of their pain with reductions in mean attack frequency, duration, intensity, and pain-related impairment scores [7]. Patients' mental health improved with reductions in depressive and anxiety related symptoms and headache medication consumption [7].

The first report and considered to be an early proof of concept for RCM management came from a case series with 17 patients who were implanted with a high-cervical SCS [9]. Over 71% of patients had >50% of pain relief including a 60% mean reduction in pain intensity ($P < .0001$) at 15 months [9]. Furthermore, there was a decrease from 28 to nine median number of migraine days ($P = .0313$), quality of life improvement, and increase in the percentage of patients not requiring additional pain medication from 0% to 37% [9]. Complications were consistent with the rate of complications for other indications for SCS indications with three patients having infections and three with lead dislocations resulting in an overall complication rate of 17.6% [9,47–51]. In a second case series, four patients with chronic migraine were implanted with a high-cervical SCS using 10 kHz, and at 28 months follow-up, all patients had at least >50% reduction in headache frequency and intensity [14].

A prospective, open-label study involving 17 patients with RCM to conservative medical management including onabotulinumtoxinA injections demonstrated further evidence for the efficacy of SCS in RCM [11]. Of the 17 patients who underwent the trial, 14 patients proceeded with the permanent high-cervical SCS placement using 10 kHz with seven patients experiencing a >30% reduction in headache days; five of these patients experienced >50% reduction, and eight reverted to an episodic pattern [11]. Moreover, migraine disability assessment score (MIDAS) and headache impact test 6 (HIT-6) questionnaires demonstrated a reduction in the headache impact and function on the quality of life and function [11]. Similar to the other aforementioned studies, this study was not without complications but was limited to three patients who had IPG/connection site tenderness and one patient requiring revision due to lead migration [11]. Though the complication rate was low and consistent with prior SCS studies, this study still had an overall complication rate lower than that of ONS [11,52,53].

A second prospective open-label study with patients with RCM was undertaken using a high-cervical SCS using 10 kHz to assess the long-term efficacy and safety for the use in these patients [12]. 20 patients with RCM were enrolled; of note, these patients had an average number of 12.2 ± 3.1 failed preventative treatments [12]. The SCS leads were inserted and advanced until the distal lead tip reached the C2 level [12]. Compared with the preimplantation period at the 52 week follow up, mean monthly migraine days decreased by 9.3 days ($P < .001$) with nearly 50% of patients converting to an episodic pattern [12]. When assessing the headache-related quality of life, there was a mean gain of 24.9 ± 23.1 ($P < .001$) on the Minnesota Satisfaction Questionnaire (MSQ), and the classification of patients considered to have severe headache disability on the HIT-6 declined to 60% from 100% [12]. Despite the presence of adverse events similar to other studies, no SCS required surgical revision [12].

Given the first reports of SCS involved in CCH patients, a systematic review was performed examining the role of cervical spinal cord stimulation in headache disorders overall [7,8,10]. In total, 16 studies involving 107 patients with disorders such as cervicogenic headache, cluster headache, migraine headache with or without aura, occipital neuralgia trigeminal neuropathy, poststroke facial pain, posttraumatic headache, and short-lasting unilateral neuralgiform headache with autonomic symptoms were included in the study [10]. Results were limited due to the studies available, which mostly consisted of case series, case reports, and observational studies, but it did reveal very low-quality evidence for a decrease in migraine headache frequency and intensity and trigeminal neuropathy intensity [10]. Of the 107 patients, only 11 patients necessitated removal suggesting the overall safety of SCS [10]. Despite the very low level of evidence, more double-blinded randomized controlled trials are needed to fully appreciate the effect of SCS on each of these as conditions evaluated in the systematic review as none exist to date.

CONCLUSION

As more data is generated on the use of SCS for RCM, several challenges lie ahead for SCS and other neuromodulation therapies for RCM. Though it is standard of practice to perform a trial of stimulation prior to permanent implantation, consideration of only proceeding to permanent implantation without a trial should be an option as a large proportion of patients in some studies proceed to permanent implant. In this regard, it would also reduce the overall cost and burden to the patient and system. Nevertheless, emerging data suggest a trial may not be necessary [11,39,54–56]. Further studies with RCM patients are needed to further assess which patients or characteristics/factors may obviate the need for a trial, proceeding directly to the permanent implant reducing the number of procedures on these patients.

Given the efficacy of ONS and other minimally invasive treatment options, determining the optimal location of SCS within the therapeutic migraine interventional management treatment ladder would greatly benefit many patients [9,46,57–61]. Future studies are warranted including double-blinded, randomized sham-controlled trials with proper experimental design that are sufficiently powered to elucidate any possible side effects are needed. In summary, a better understanding of clinical efficacy and safety profile will potentially provide greater clarity on the mechanisms of action for SCS in the treatment of RCM. Addressing these challenges is crucial for the advancement of the field of neuromodulation in RCM.

DISCLAIMER

The content within this chapter is for educational and not promotional purposes. SCS for RCM remains off label and investigational as of the writing of this chapter.

REFERENCES

[1] D.M. Moore, C. McCrory, Spinal cord stimulation, BJA Educ. 16 (8) (2016) 258–263, https://doi.org/10.1093/bjaed/mkv072.
[2] T.J. Lamer, S.M. Moeschler, H.M. Gazelka, W.M. Hooten, M.A. Bendel, M.H. Murad, Spinal stimulation for the treatment of intractable spine and limb pain: a systematic review of RCTs and meta-analysis, Mayo Clin. Proc. 94 (8) (2019) 1475–1487, https://doi.org/10.1016/j.mayocp.2018.12.037.
[3] P.L. Gildenberg, History of electrical neuromodulation for chronic pain, Pain Med. 7 (1) (2006) S7, https://doi.org/10.1111/j.1526-4637.2006.00118.x.
[4] E.A. Petersen, T.G. Stauss, J.A. Scowcroft, E.S. Brooks, J.L. White, S.M. Sills, K. Amirdelfan, M.N. Guirguis, J. Xu, C. Yu, A. Nairizi, D.G. Patterson, K.C. Tsoulfas, M.J. Creamer, V. Galan, R.H. Bundschu, C.A. Paul, N.D. Mehta, H. Choi, D. Sayed, S.P. Lad, D.J. Dibenedetto, K.A. Sethi, J.H. Goree, M.T. Bennett, N.J. Harrison, A.F. Israel, P. Chang, P.W. Wu, G. Gekht, C.E. Argoff, C.E. Nasr, R.S. Taylor, J. Subbaroyan, B.E. Gliner, D.L. Caraway, N.A. Mekhail, Effect of high-frequency (10-kHz) spinal cord stimulation in patients with painful diabetic neuropathy: a randomized clinical trial, JAMA Neurol. 78 (6) (2021) 687–698, https://doi.org/10.1001/jamaneurol.2021.0538.
[5] P. Verrills, C. Sinclair, A. Barnard, A review of spinal cord stimulation systems for chronic pain, J. Pain Res. 9 (2016) 481–492, https://doi.org/10.2147/JPR.S108884.
[6] Abbott Receives FDA Approval for its Spinal Cord Stimulation Systems to Treat Chronic Back Pain in People Who Have Limited Surgical Options.
[7] T. Wolter, A. Kiemen, H. Kaube, High cervical spinal cord stimulation for chronic cluster headache, Cephalalgia 31 (11) (2011) 1170–1180, https://doi.org/10.1177/0333102411412627.
[8] T. Wolter, H. Kaube, M. Mohadjer, High cervical epidural neurostimulation for cluster headache: case report and review of the literature, Cephalalgia 28 (10) (2008) 1091–1094, https://doi.org/10.1111/j.1468-2982.2008.01661.x.
[9] R. De Agostino, B. Federspiel, E. Cesnulis, P.S. Sandor, High-cervical spinal cord stimulation for medically intractable chronic migraine, Neuromodulation 18 (4) (2015) 289–296, https://doi.org/10.1111/ner.12236Switzerland.
[10] M.T. Finnern, R.S. D'Souza, M.Y. Jin, A.A. Abd-Elsayed, Cervical spinal cord stimulation for the treatment of headache disorders: a systematic review, Neuromodulation 26 (7) (2023) 1309–1318, https://doi.org/10.1016/j.neurom.2022.10.060.
[11] R. Arcioni, S. Palmisani, M. Mercieri, V. Vano, S. Tigano, T. Smith, M.R.D. Fiore, A. Al-Kaisy, P. Martelletti, Cervical 10 kHz spinal cord stimulation in the management of chronic, medically refractory migraine: a prospective, open-label, exploratory study, Eur. J. Pain 20 (1) (2016) 70–78, https://doi.org/10.1002/ejp.692.
[12] A. Al-Kaisy, S. Palmisani, R. Carganillo, S. Wesley, D. Pang, A. Rotte, A. Santos, G. Lambru, Safety and efficacy of 10 kHz spinal cord stimulation for the treatment of refractory chronic migraine: a prospective long-term open-label study, Neuromodulation 25 (1) (2022) 103–113, https://doi.org/10.1111/ner.13465.
[13] H. Yuan, T.Y. Chuang, Update of neuromodulation in chronic migraine, Curr. Pain Headache Rep. 25 (11) (2021), https://doi.org/10.1007/s11916-021-00988-7.
[14] G. Lambru, M. Trimboli, S. Palmisani, T. Smith, A. Al-Kaisy, Safety and efficacy of cervical 10 kHz spinal cord stimulation in chronic refractory primary headaches: a retrospective case series, J. Headache Pain 17 (1) (2016), https://doi.org/10.1186/s10194-016-0657-2.
[15] A.D. Sdrulla, Y. Guan, S.N. Raja, Spinal cord stimulation: clinical efficacy and potential mechanisms, Pain Pract. 18 (8) (2018) 1048–1067, https://doi.org/10.1111/papr.12692.
[16] I.E. Harmsen, D. Hasanova, G.J.B. Elias, A. Boutet, C. Neudorfer, A. Loh, J. Germann, A.M. Lozano, Trends in clinical trials for spinal cord stimulation, Stereotact. Funct. Neurosurg. 99 (2) (2021) 123–134, https://doi.org/10.1159/000510775.

[17] L. Kapural, C. Yu, M.W. Doust, B.E. Gliner, R. Vallejo, B. Todd Sitzman, K. Amirdelfan, D.M. Morgan, L.L. Brown, T.L. Yearwood, R. Bundschu, A.W. Burton, T. Yang, R. Benyamin, A.H. Burgher, Novel 10-kHz high-frequency therapy (HF10 therapy) is superior to traditional low-frequency spinal cord stimulation for the treatment of chronic back and Leg pain, Anesthesiology 123 (4) (2015) 851–860, https://doi.org/10.1097/aln.0000000000000774.

[18] C.C. De Vos, M.J. Bom, S. Vanneste, M.W.P.M. Lenders, D. De Ridder, Burst spinal cord stimulation evaluated in patients with failed back surgery syndrome and painful diabetic neuropathy, Neuromodulation 17 (2) (2014) 152–159, https://doi.org/10.1111/ner.12116.

[19] D. De Ridder, S. Vanneste, M. Plazier, E. Van Der Loo, T. Menovsky, Burst spinal cord stimulation: toward paresthesia-free pain suppression, Neurosurgery 66 (5) (2010) 986–990, https://doi.org/10.1227/01.NEU.0000368153.44883.B3.

[20] A. Crossman, D. Neary, Spinal Cord, 2020, p. 2020.

[21] R.M. Levy, Anatomic considerations for spinal cord stimulation, Neuromodulation 17 (1) (2014) 2–11, https://doi.org/10.1111/ner.12175.

[22] R. Melzack, P.D. Wall, Pain mechanisms: a new theory, Science 150 (3699) (1965) 971–979, https://doi.org/10.1126/science.150.3699.971.

[23] V. Yakhnitsa, B. Linderoth, B.A. Meyerson, Spinal cord stimulation attenuates dorsal horn neuronal hyperexcitability in a rat model of mononeuropathy, Pain 79 (2–3) (1999) 223–233, https://doi.org/10.1016/S0304-3959(98)00169-9.

[24] B. Linderoth, B. Gazelius, J. Franck, E. Brodin, Dorsal column stimulation induces release of serotonin and substance P in the cat dorsal horn, Neurosurgery 1 (2) (1992) 289–297, https://doi.org/10.1227/00006123-199208000-00014.

[25] B. Linderoth, C.O. Stiller, L. Gunasekera, W.T. O'Connor, U. Ungerstedt, E. Brodin, Gamma-aminobutyric acid is released in the dorsal horn by electrical spinal cord stimulation: an in vivo microdialysis study in the rat, Neurosurgery 34 (3) (1994) 484–489, https://doi.org/10.1227/00006123-199403000-00014.

[26] J. Barchini, S. Tchachaghian, F. Shamaa, S.J. Jabbur, B.A. Meyerson, Z. Song, B. Linderoth, N.E. Saadé, Spinal segmental and supraspinal mechanisms underlying the pain-relieving effects of spinal cord stimulation: an experimental study in a rat model of neuropathy, Neuroscience 215 (2012) 196–208, https://doi.org/10.1016/j.neuroscience.2012.04.057.

[27] G. Schechtmann, Z. Song, C. Ultenius, B.A. Meyerson, B. Linderoth, Cholinergic mechanisms involved in the pain relieving effect of spinal cord stimulation in a model of neuropathy, Pain 139 (1) (2008) 136–145, https://doi.org/10.1016/j.pain.2008.03.023.

[28] C.O. Stiller, J.G. Cui, W.T. O'Connor, E. Brodin, B.A. Meyerson, B. Linderoth, Release of γ-aminobutyric acid in the dorsal horn and suppression of tactile allodynia by spinal cord stimulation in mononeuropathic rats, Neurosurgery 39 (2) (1996) 367–375, https://doi.org/10.1097/00006123-199608000-00026.

[29] K.F. McCarthy, T.J. Connor, C. McCrory, Cerebrospinal fluid levels of vascular endothelial growth factor correlate with reported pain and are reduced by spinal cord stimulation in patients with failed back surgery syndrome, Neuromodulation 16 (6) (2013) 519–522, https://doi.org/10.1111/j.1525-1403.2012.00527.x.

[30] T. Bartsch, P.J. Goadsby, Stimulation of the greater occipital nerve induces increased central excitability of dural afferent input, Brain 125 (7) (2002) 1496–1509, https://doi.org/10.1093/brain/awf166.

[31] T. Bartsch, P.J. Goadsby, Increased responses in trigeminocervical nociceptive neurons to cervical input after stimulation of the dura mater, Brain 126 (8) (2003) 1801–1813, https://doi.org/10.1093/brain/awg190.

[32] A. Ambrosini, M. Vandenheede, P. Rossi, F. Aloj, E. Sauli, F. Pierelli, J. Schoenen, Suboccipital injection with a mixture of rapid- and long-acting steroids in cluster headache: a double-blind placebo-controlled study, Pain 118 (1) (2005) 92–96, https://doi.org/10.1016/j.pain.2005.07.015.

[33] K.Y. Lee, C. Bae, D. Lee, Z. Kagan, K. Bradley, J.M. Chung, J.H. La, Low-intensity, kilohertz frequency spinal cord stimulation differently affects excitatory and inhibitory

neurons in the rodent superficial dorsal horn, Neuroscience 428 (2020) 132−139, https://doi.org/10.1016/j.neuroscience.2019.12.031.

[34] J.C. Rains, J.S. Poceta, Sleep-related headaches, Neurol. Clin. 30 (4) (2012) 1285−1298, https://doi.org/10.1016/j.ncl.2012.08.014.

[35] H. Saçmacı, N. Tanik, L.E. İnan, Current perspectives on the impact of chronic migraine on sleep quality: a literature review, Nat. Sci.Sleep 14 (2022) 1783−1800, https://doi.org/10.2147/NSS.S335949.

[36] S. Duan, Z. Ren, H. Xia, Z. Wang, T. Zheng, Z. Liu, Association between sleep quality, migraine and migraine burden, Front. Neurol. 13 (2022), https://doi.org/10.3389/fneur.2022.955298.

[37] C. Tiseo, A. Vacca, A. Felbush, T. Filimonova, A. Gai, T. Glazyrina, I.A. Hubalek, Y. Marchenko, L. Hendrik Overeem, S. Piroso, A. Tkachev, P. Martelletti, S. Sacco, Migraine and sleep disorders: a systematic review, J. Headache Pain 21 (1) (2020), https://doi.org/10.1186/s10194-020-01192-5.

[38] M. Waliszewska-Prosół, M. Nowakowska-Kotas, J. Chojdak-łukasiewicz, S. Budrewicz, Migraine and sleep—an unexplained association? Int. J. Mol. Sci. 22 (2021) 2021.

[39] S. Palmisani, A. Al-Kaisy, R. Arcioni, T. Smith, A. Negro, G. Lambru, V. Bandikatla, E. Carson, P. Martelletti, A six year retrospective review of occipital nerve stimulation practice–controversies and challenges of an emerging technique for treating refractory headache syndromes, J. Headache Pain 14 (2013) 67, https://doi.org/10.1186/1129-2377-14-67.

[40] S. Miller, L. Watkins, M. Matharu, Long-term outcomes of occipital nerve stimulation for chronic migraine: a cohort of 53 patients, J. Headache Pain 17 (1) (2016), https://doi.org/10.1186/s10194-016-0659-0.

[41] A.C. Brewer, T.L. Trentman, M.G. Ivancic, B.B. Vargas, A.M. Rebecca, R.S. Zimmerman, D.M. Rosenfeld, D.W. Dodick, Longa-term outcome in occipital nerve stimulation patients with medically intractable primary headache disorders, Neuromodulation 16 (6) (2013) 557−564, https://doi.org/10.1111/j.1525-1403.2012.00490.x.

[42] P. Notaro, The effects of peripheral occipital nerve stimulation for the treatment of patients suffering from chronic migraine: a single center experience.

[43] K. Paemeleire, J.P. Van Buyten, M. Van Buynder, D. Alicino, G. Van Maele, I. Smet, P.J. Goadsby, Phenotype of patients responsive to occipital nerve stimulation for refractory head pain, Cephalalgia 30 (6) (2010) 662−673, https://doi.org/10.1111/j.1468-2982.2009.02022.x.

[44] C.A. Popeney, K.M. Aló, Peripheral neurostimulation for the treatment of chronic, disabling transformed migraine, Headache 43 (4) (2003) 369−375, https://doi.org/10.1046/j.1526-4610.2003.03072.x.

[45] Clinico Universitario Lozano Blesa. Occipital Nerve Stimulation for Refractory Chronic Migraine: Results of a Long-Term Prospective Study.

[46] J.R. Saper, D.W. Dodick, S.D. Silberstein, S. McCarville, M. Sun, P.J. Goadsby, Occipital nerve stimulation for the treatment of intractable chronic migraine headache: ONSTIM feasibility study, Cephalalgia 31 (3) (2011) 271−285, https://doi.org/10.1177/0333102410381142.

[47] M.A. Bendel, T. O'Brien, B.C. Hoelzer, T.R. Deer, T.P. Pittelkow, S. Costandi, D.R. Walega, G. Azer, S.M. Hayek, Z. Wang, J.S. Eldrige, W. Qu, J.M. Rosenow, S.M. Falowski, S.A. Neuman, S.M. Moeschler, C. Wassef, C. Kim, T. Niazi, T. Saifullah, B. Yee, C. Kim, C.L. Oryhan, D.T. Warren, I. Lerman, R. Mora, M. Hanes, T. Simopoulos, S. Sharma, C. Gilligan, W. Grace, T. Ade, N.A. Mekhail, J.P. Hunter, D. Choi, D.Y. Choi, Spinal cord stimulator related infections: findings from a multicenter retrospective analysis of 2737 implants, Neuromodulation 20 (6) (2017) 553−557, https://doi.org/10.1111/ner.12636.

[48] T. Cameron, Safety and efficacy of spinal cord stimulation for the treatment of chronic pain: a 20-year literature review, J. Neurosurg. Spine 100 (3) (2004) 254−267, https://doi.org/10.3171/spi.2004.100.3.0254.

[49] K. Kumar, G. Hunter, D. Demeria, Spinal cord stimulation in treatment of chronic benign pain: challenges in treatment planning and present status, a 22-year experience, Neurosurgery 58 (3) (2006) 481–494, https://doi.org/10.1227/01.NEU.0000192162.99567.96.

[50] D.G.A.J. Quigley, P.R. Eldridge, H. Cameron, K. McIvor, J.B. Miles, T.R.K. Varma, Long-term outcome of spinal cord stimulation and hardware complications, Stereotact. Funct. Neurosurg. 81 (1–4) (2003) 50–56, https://doi.org/10.1159/000075104.

[51] B.C. Hoelzer, M.A. Bendel, T.R. Deer, J.S. Eldrige, D.R. Walega, Z. Wang, S. Costandi, G. Azer, W. Qu, S.M. Falowski, S.A. Neuman, S.M. Moeschler, C. Wassef, C. Kim, T. Niazi, T. Saifullah, B. Yee, C. Kim, C.L. Oryhan, J.M. Rosenow, D.T. Warren, I. Lerman, R. Mora, S.M. Hayek, M. Hanes, T. Simopoulos, S. Sharma, C. Gilligan, W. Grace, T. Ade, N.A. Mekhail, J.P. Hunter, D. Choi, D.Y. Choi, Spinal cord stimulator implant infection rates and risk factors: a multicenter retrospective study, Neuromodulation 20 (6) (2017) 558–562, https://doi.org/10.1111/ner.12609.

[52] O. Mueller, H.-C. Diener, P. Dammann, K. Rabe, H. Vincent, S. Ulrich, C. Gaul, Occipital nerve stimulation for intractable chronic cluster headache or migraine: a critical analysis of direct treatment costs and complications, Cephalalgia 33 (16) (2013) 1283–1291, https://doi.org/10.1177/0333102413493193.

[53] D.W. Dodick, S.D. Silberstein, K.L. Reed, T.R. Deer, K.V. Slavin, B. Huh, A.D. Sharan, S. Narouze, A.Y. Mogilner, T.L. Trentman, J. Ordia, J. Vaisman, J. Goldstein, N. Mekhail, Safety and efficacy of peripheral nerve stimulation of the occipital nerves for the management of chronic migraine: long-term results from a randomized, multicenter, double-blinded, controlled study, Cephalalgia 35 (4) (2015) 344–358, https://doi.org/10.1177/0333102414543331.

[54] R.V. Duarte, R. Houten, S. Nevitt, M. Brookes, J. Bell, J. Earle, A. Gulve, S. Thomson, G. Baranidharan, R.B. North, R.S. Taylor, S. Eldabe, Screening trials of spinal cord stimulation for neuropathic pain in England—a budget impact analysis, Front. Pain Res. 3 (2022), https://doi.org/10.3389/fpain.2022.974904.

[55] S. Eldabe, S. Nevitt, S. Griffiths, A. Gulve, S. Thomson, B. Ganesan, R. Houten, M. Brookes, A. Kansal, J. Earle, J. Bell, R.S. Taylor, R.V. Duarte, Does a screening trial for spinal cord stimulation in patients with chronic pain of neuropathic origin have clinical utility (TRIAL-STIM)? 36-Month results from a randomized controlled trial, Neurosurgery 92 (1) (2023) 75–82, https://doi.org/10.1227/neu.0000000000002165.

[56] R. Chadwick, R. McNaughton, S. Eldabe, G. Baranidharan, J. Bell, M. Brookes, R.V. Duarte, J. Earle, A. Gulve, R. Houten, S. Jowett, A. Kansal, S. Rhodes, J. Robinson, S. Griffiths, R.S. Taylor, S. Thomson, H. Sandhu, To trial or not to trial before spinal cord stimulation for chronic neuropathic pain: the patients' view from the TRIAL-STIM randomized controlled trial, Neuromodulation 24 (3) (2021) 459–470, https://doi.org/10.1111/ner.13316.

[57] G. Serra, F. Marchioretto, Occipital nerve stimulation for chronic migraine: a randomized trial, Pain Physician 15 (3) (2012) 245–253.

[58] R.H. Dworkin, D.C. Turk, K.W. Wyrwich, D. Beaton, C.S. Cleeland, J.T. Farrar, J.A. Haythornthwaite, M.P. Jensen, R.D. Kerns, D.N. Ader, N. Brandenburg, L.B. Burke, D. Cella, J. Chandler, P. Cowan, R. Dimitrova, R. Dionne, S. Hertz, A.R. Jadad, N.P. Katz, H. Kehlet, L.D. Kramer, D.C. Manning, C. McCormick, M.P. McDermott, H.J. McQuay, S. Patel, L. Porter, S. Quessy, B.A. Rappaport, C. Rauschkolb, D.A. Revicki, M. Rothman, K.E. Schmader, B.R. Stacey, J.W. Stauffer, T. von Stein, R.E. White, J. Witter, S. Zavisic, Interpreting the clinical importance of treatment outcomes in chronic pain clinical trials: IMMPACT recommendations, J. Pain 9 (2) (2008) 105–121, https://doi.org/10.1016/j.jpain.2007.09.005.

[59] S.D. Silberstein, D.W. Dodick, J. Saper, B. Huh, K.V. Slavin, A. Sharan, K. Reed, S. Narouze, A. Mogilner, J. Goldstein, T. Trentman, J. Vaisman, J. Ordia, P. Weber, T. Deer, R. Levy, R.L. Diaz, S.N. Washburn, N. Mekhail, Safety and efficacy of peripheral nerve stimulation of the occipital nerves for the management of chronic migraine:

results from a randomized, multicenter, double-blinded, controlled study, Cephalalgia 32 (16) (2012) 1165–1179, https://doi.org/10.1177/0333102412462642.

[60] D.J. Ellens, R.M. Levy, Peripheral neuromodulation for migraine headache, Prog. Neurol. Surg. 24 (2011) 109–117, https://doi.org/10.1159/000323890.

[61] P.J. Goadsby, T. Sprenger, Current practice and future directions in the prevention and acute management of migraine, Lancet Neurol. 9 (3) (2010) 285–298, https://doi.org/10.1016/S1474-4422(10)70005-3.

Chapter | Fourteen

Transcranial magnetic stimulation (TMS)

Anthony J. Batri[1], Crystal Joseph[2], Tsan-Chen Yeh[3], Roshan Santhosh[4]

[1]New York-Presbyterian Hospital, the University Hospitals of Columbia and Cornell, Columbia University Irving Medical Center, New York, NY, United States; [2]Beth Israel Deaconess Medical Center, Boston, MA, United States; [3]Riverside Community Hospital, Riverside, CA, United States; [4]Larkin Community Hospital, South Miami, FL, United States

INTRODUCTION

Transcranial magnetic stimulation (TMS) is a noninvasive technique that targets specific brain regions implicated for pain processing, potentially disrupting the abnormal pain signaling pathway. In this chapter, we discuss the potential of TMS in treatment of various headache disorders.

Basics of transcranial magnetic stimulation (TMS)

TMS creates rapidly alternating magnetic fields and forms an electric current, which can be induced in a nearby conductor. Depending on the type of neuron targeted, this current may excite or inhibit the area of the brain below the coil.

The utility of TMS in headache management

While most literature focused on TMS's therapeutic effect in migraine management, potential TMS's utilization has also been investigated in treatment of other headache disorders, including but not limited to: tension-type headache, medication overuse headache, posttraumatic headache, cluster headache, and facial pain.

Safety and adverse effects

TMS is generally well tolerated by patients with minimal adverse effects. Common adverse effects observed with TMS include headache, scalp/neck pain, sinusitis, and numbness. Other more rare side adverse events include seizure induction and potential hearing loss.

Interventional Management of Migraines and Other Headache Disorders. https://doi.org/10.1016/B978-0-443-23557-3.00014-0

Patient outcomes and satisfaction

After TMS treatment, patients generally saw an improvement in symptoms as measured by pain severity and duration of symptom resolution after treatment.

INDICATIONS AND CONTRAINDICATIONS TO CONSIDER IN PATIENT SELECTION FOR TMS

For providers, there are important indications and contraindications to consider for patients regarding TMS treatment. TMS is approved for prophylactic and abortive migraine treatment but can also be used for patients who are 12 years old and older based on current literature. It is important to consider the contraindications for TMS, such as the risk of seizures, metal or electronic implants, implanted medical devices, and cochlear implants that can be close to the TMS coil magnetic fields.

The role of TMS compared to other neuromodulation techniques for headaches

Neuromodulation is defined as an intervention that either potentiates or inhibits nerve impulse transmission. For migraine treatment, there are many neuromodulation techniques available. Looking at the type of stimulation, target, practical use, and CE or FDA approval is important to determine which technique is optimal for the patient.

Future considerations in the use of TMS for headaches

As an FDA and CE approved intervention for acute and prophylactic migraines, TMS has excellent potential in migraine treatment. However, the literature regarding TMS varies in quality of evidence. Larger and better-designed trials are needed before TMS is widely used as part of mainstay clinical care.

INTRODUCTION

Transcranial magnetic stimulation (TMS) is a noninvasive technique that offers both diagnostic and therapeutic benefits to a wide variety of neurological and psychiatric disorders. In recent years, increasing evidence has emerged and encouraged increased investigation on TMS efficacy in managing various types of pain, including headache disorders [1].

By generating a brief, high-intensity magnetic field, TMS applied fluctuating magnetic pulses to specific areas of the brain. In addition to mapping brain function and exploring the excitability of different regions, including those responsible for sensory, motor, and cognitive function, in recent years multiple studies have been conducted to evaluate its therapeutic utility in pain management. By targeting specific brain regions implicated for pain processing, TMS can potentially disrupt the abnormal pain signaling pathway [2].

While the underlying mechanisms of TMS in headache management are not yet fully understood, the evidence for TMS includes randomized controlled trials for various headache disorders, with most studies done in patients with migraines with aura. It is proposed that stimulation of cortical region by the magnetic pulses can lead to sustained excitation of a brain region and release of neurotransmitters that are involved in modulation of pain [3]. In this chapter, we will discuss the potential role of TMS in treatment of various headache disorders, which could provide more options when it comes to targeted and personalized pain management plans.

BASICS OF TMS

TMS applies Faraday's law of electromagnetic induction, which states that, by rapidly alternating magnetic fields, an electric current can be induced in a nearby conductor. During TMS, a brief electric current is generated through a magnetic coil, which induces a high intensity magnetic field (up to 2 T) [2]. When applied in a rapidly alternating manner, an electric current in the brain is induced parallel to the plane of the magnetic coil. Depending on the type of neuron targeted, this current may excite or inhibit the area of the brain below the coil by causing neuronal depolarization [2]. There are various protocols of TMS, including single pulse TMS (sTMS), paired-pulse TMS, repetitive TMS (rTMS), and deep TMS. Each protocol can be used for different indications.

sTMS has been used to determine neuronal thresholds, or the intensity required to elicit a chain of action potentials in the targeted cortical region. For example, a phosphene, or a phenomenon of seeing light without any light actually entering the eye, is produced when TMS is applied over the visual cortex. It has been proposed that patients with migraines are more likely to have a lower phosphene threshold than control participants [4].

In rTMS, a series of TMS pulses is applied at frequencies ranging from 1 to 50Hz. In a study done by Rossi et al. [5], low frequency rTMS (1Hz) has been shown to inhibit cortical excitability, while higher frequency stimulation at 5–20 Hz has been shown to increase cortical excitability [5]. Due to its long lasting effects, the efficacy of rTMS has been investigated in treating various neurologic and psychiatric conditions, including depression, anxiety, cerebral palsy, neurodegenerative disease, headaches, etc.

Evidence of TMS in headache treatments

While most literature focused on TMS's therapeutic effect in migraine treatment, potential TMS's utilization has also been investigated in treatment of other headache disorders, including but not limited to: tension-type headache, medication overuse headache, posttraumatic headache, cluster headache, and facial pain.

Migraine

STMS FOR ACUTE TREATMENT OF MIGRAINE

sTMS is presumed to abort migraine attacks with aura through inhibition of cortical spreading depression (CSD). CSD is believed to be one of the causes for migraine aura [3]. (Lipton and Pearlman) Most common form of CSD occurs in the occipital cortex, giving rise to visual aura. Efficacy of sTMS as an acute treatment has then been studied and reported in several studies, first with large, tabletop devices.

In a small pilot study, 42 subjects with migraine (5 with aura) were randomized to receive either high intensity or low intensity double-pulse sTMS. During onset of their migraine attacks, subjects were instructed to go to the clinic where sTMS was administered with a tabletop device. The headache pain was then monitored at 5-minute intervals for 20 minutes or less after treatment. 75% decrease in headache intensity from baseline was reported. 32% of those who received one treatment (n = 42 and with a maximum of three sTMS stimulation), reported no further headache at 24 hours. Furthermore, for patients with aura in the study (n = 5), all of them achieved pain relief with sTMS treatment [6].

To address issues of large tabletop TMS devices, which include having to travel to the clinic, which can delay treatment and potentially diminish TMS's therapeutic efficacy, a lightweight and portable TMS device was introduced. Lipton et al. investigated the effect of TMS for acute treatment of migraine with aura with these portable devices. In this multicenter, randomized, double blind, sham-control study, sTMS was delivered to the occipital cortex through a self-monitored device. Subjects were instructed to treat as soon as possible after migraine aura began within 1 hour of onset. The primary efficacy outcome was the 2-hour pain free response. The result shows that early treatment of migraine with aura with sTMS resulted in significantly higher pain free rate. The absence of migraine pain was also sustained 24 and 48 hours after treatment [7].

It should be noted that additional studies are still needed for sTMS in treatment of migraine without aura. The mechanism of CSD in migraine without aura remains unclear. However, in a study done by Goadsby et al. [8], CSD in cortex has been associated with the premonitory phase of migraine, or the unpainful symptoms migraine patients experience days before migraine episode [8]. The efficacy of sTMS should be studied in patients without aura with and without a premonitory phase.

RTMS AS A PREVENTATIVE TREATMENT OF MIGRAINE

Based on the same principle that sTMS might treat migraine acutely by disrupting CSD, it has been proposed that rTMS may be helpful in migraine prevention. Multiple studies have tested the efficacy of rTMS in migraine prophylaxis. In a small pilot study, Brighina et al. [9] evaluate the therapeutic effect of high-frequency rTMS over left dorsolateral prefrontal cortex (DLPFC) in relieving chronic migraine. Frequency of migraine attack, headache index, and number of acute medications used were recorded and compared between the treatment group (n = 6) and placebo (n = 5). Subjects treated by rTMS showed a significant improvement over baseline in attack frequency and acute medication/treatment use [9]. In another study, Teepker et al. [10] investigated the efficacy of low-frequency rTMS over vertex in migraine prevention. In this placebo-controlled, blinded study, the primary end-point was defined as migraine attack compared with placebo, while the secondary outcomes were reduction in the total days and hours, pain intensity, and decrease of analgesic intake for migraine. While no significance was evident in the result, decreased frequency of migraine attacks, days, and duration was observed [10].

The positive prophylactic effects of rTMS that we observed so far are encouraging. While the exact mechanism of rTMS as a preventive treatment for migraine still needs to be further investigated, further research on this topic is indicated. Aside from producing sustained changes in brain excitability, the role of rTMS in modulating neurotransmitter levels, including dopamine and glutamine/glutamate [11–13], should also be further studied for its therapeutic effect.

Chronic tension type headache

Efficacy of TMS in chronic tension-type headache (CTTH) has also been investigated. In a study done by Kalita et al., efficacy of three versus one session of 10 Hz rTMS is assessed in both chronic migraine (CM) and CTTH subjects. The primary outcome was 50% reduction in headache frequency while secondary outcomes were improvement in headache severity, functional disability, abortive medication use, and side effects. Significant decrease in both primary and secondary outcomes was observed in all subjects, regardless of whether they receive one or three sessions of rTMS. Conversion of chronic to episodic headache in both CM and CTTH subjects was also observed [14].

POST TRAUMATIC HEADACHE

Being one of the various postconcussive symptoms (PCS), posttraumatic headache can also interfere with daily activity of TBI patients. As subjects with post traumatic headache usually respond poorly to pharmacotherapy, efficacy of TMS in post traumatic headache has been investigated.

In a study done by Koski et al. [15], rTMS over the DLPFC was tested for alleviating PCS in subjects after mild TBI. Primary outcomes of the study include tolerability, safety, and efficacy, as measured with the PCS Scale. At 3 months follow up, subjects who completed the study all reported reduction in PCS severity. In addition, some participants also report less sleep disturbance (n = 3) and better mental focus (n = 3) [15]. In another study done by Leung et al. [16], efficacy of short course rTMS was evaluated at treating persistent mild TBI related headache in addition to associated neuropsychological dysfunction after TBI. The result shows 50% and 57% reduction in prevalence of persistent headache at 1 and 4 week post treatment in subjects who completed the study [16]. While further studies are required to establish a clinical protocol balancing both efficacy and patient compliance, the role of TMS in posttraumatic headache management should not be undermined.

Other headache disorders

While the number of studies is limited, efficacy of TMS has also been investigated in other headache disorders. In a study done by Hodag et al. [17], the long-term therapeutic effect of high frequency rTMS over the motor cortex contralateral to pain is assessed in patients with cluster headache, trigeminal neuropathic pain, and atypical facial pain. In the study, rTMS protocol consisted of an induction phase, which consisted of one daily rTMS session for 5 days weekly for two consecutive weeks followed by a maintenance phase of two sessions during 1 week, then one session in week four and 6, and a monthly session for the next 5 months. In a subset of subjects, sessions of rTMS are shortened to 10 minutes from 20 minutes with the same number of pulses. All measures of pain, including permanent pain intensity, paroxysmal pain, and number of daily attacks, show significant decrease. This study shows that a long-term protocol of rTMS consisting of induction and maintenance phase could be a potential therapeutic option in pain management of patients with chronic facial pain and cluster headache [17]).

INDICATIONS AND CONTRAINDICATIONS TO CONSIDER IN PATIENT SELECTION FOR TMS

For providers, there are important considerations for optimal patient selection regarding TMS. A comprehensive evaluation of the patient's medical information (medical history, surgical history, neurological history, psychiatric history, and medications) and physical examination is vital to determine the necessity of TMS and its safety.

TMS is clinically indicated for a variety of pain pathologies beyond headache disorders. For headache disorders, TMS was approved in 2014 for prophylactic and abortive migraine treatment. All patients with a confirmed diagnosis of migraine can be considered for TMS. TMS can be used alone or in conjunction with pharmacotherapy [18]. Neuromodulation techniques such as TMS can be considered for acute or preventive treatment of migraines when conservative treatments have been trialed with inadequate response. TMS can be used for patients who are 12 years old and older based on the literature. For patients suffering from frequent attacks, this novel approach can be used to decrease medication use, subsequently preventing medication-overuse headache [18]. Indications for preventive treatment include episodes that significantly interfere with quality of life despite acute treatment, frequent attacks, and failure, overuse or contraindication of acute treatments. Overuse is defined as either 10 or more days per month of using ergot-derived medications, opioids, multiple drugs from different classes, and triptans or 15 or more days per month of non-opioid analgesics, nonsteroidal antiinflammatory drugs, and acetaminophen

[18]. Patients who prefer to avoid medication or have poor tolerability or contraindications to medications can be considered for TMS. For preventive treatment, a trial of TMS should be done concurrently with the established treatment regimen. Additionally, providers should individualize the treatment plan, tailoring the overall regimen based on the patient.

When determining the treatment plan, it is important to consider the contraindications for TMS. A major contraindication for TMS is the risk of seizures, which is the most common complication associated with the procedure [19]. In patients without seizure disorder, the risk of seizure is $<0.01\%$ per session. However, those who have a seizure history have a higher risk but still low, less than 3% per treatment [20]. Therefore, for patients with a seizure disorder, TMS is generally not recommended. Similarly, patients with prior neurological disease, family history of seizures, history of head trauma, previous loss of consciousness, stroke history, brain malignancy, traumatic brain injury, adolescents, medications, and substance use disorder should also be carefully considered as the seizure threshold is lower in these subset populations. For pregnant patients with headache disorders, TMS is a relatively safe procedure.

Metal or electronic implants, implanted medical devices, and cochlear implants included, that can be close to the TMS coil magnetic fields is another contraindication. Metal objects close to the TMS magnetic field can cause thermal injury to adjacent tissue, and the TMS magnetic field can cause movement of nearby metal objects [21]. As part of the comprehensive evaluation, patients should be evaluated regarding prior medical devices, exposure to metal fragments, tattoos containing ferromagnetic-containing ink, piercings, and other metal sources near the head and neck. Additionally, TMS can cause currents in subcutaneous leads in the scalp, for example, in deep brain stimulators. Inadvertent currents can flow in deep brain stimulator electrodes with TMS. Therefore, patients who have deep brain stimulators should not receive TMS until further safety testing, device modifications, and measures are taken to ensure safety.

Metal or electronic implants below the head and neck area are generally considered safe due to distance from the magnetic field. Nonferromagnetic orthodontic hardware such as braces, implants and fillings is considered safe with TMS. If a provider is unsure of the patient's clinical history and exposures, radiographs should be considered. Unfortunately, radiographic studies cannot determine if metal objects are ferromagnetic. Regarding vagal nerve stimulators and cardiac pacemakers, there is limited safety data [22]. Therefore, providers should consider consulting and discussing with other specialists prior to TMS treatment.

As part of informed consent, it is important to discuss not only the benefits but also the potential complications and ways to mitigate complications with the patient because this will impact medical decision-making and ultimately, patient selection. As previously mentioned, a seizure is the most common adverse effect. Other less severe complications include transient tinnitus, local neck pain, toothache or headache, and hyperacusis [23]. With the exception of seizures, most adverse effects are usually short-term and mild.

When there is a significant change in risk or benefit of TMS, patients should be reconsented. Within the discussion of informed consent, providers should discuss the policy regarding the number of treatments or time frame agreed upon initially, and reconsent should be done after this initial treatment period is completed. Reconsent should also be considered when transitioning from acute to maintenance TMS and changes in treatment settings.

As providers, having thorough medical decision-making discussions with patients will help to guide and provide the medical care that patients desire. Patients' wishes will certainly impact patient selection as well. For example, patients who prefer to pursue medication and other conservative management should be opted out of TMS. Lastly, the patient's fiscal coverage and follow-up ability for repeated sessions will also determine medical decision-making and therefore, patient selection. If patients are unable to pursue further treatment sessions, TMS will not be as effective.

TMS versus other neuromodulation techniques for headaches

Neuromodulation is defined as an intervention that either potentiates or inhibits nerve impulse transmission. Neuromodulation can be a safe alternative to pharmacological therapy of migraines especially in populations that are sensitive such as adolescents and pregnant women. For those who have poor tolerability for pharmacologic treatments or lack of efficacy, neuromodulation can be an excellent alternative. There are recent published literature reviews, systematic reviews and meta-analyses analyzing randomized controlled trials on neurostimulation techniques for migraine treatment. For migraine treatment, there are many neuromodulation techniques available: remote electrical neuromodulation (REN), invasive occipital nerve stimulation (ONS), noninvasive vagal nerve stimulation (nVNS), sphenopalatine ganglion stimulation (SGS), deep brain stimulation (DBS), supra-orbital transcutaneous electrical nerve stimulation (TENS), percutaneous electrical nerve stimulation (PENS), and single pulse (sp-TMS) or high-frequency repetitive transcranial magnetic stimulation (rTMS) (Fig. 14.1) [24].

Looking at the type of stimulation, target, practical use, and CE or FDA approval is important to determine which technique to use (Table 14.1), and reviewing studies major conclusions provides insight into deciding which technique is optimal for the patient. Based on a 2020 systematic review, REN appears to be effective for acute migraine treatment, and invasive occipital nerve stimulation for prophylactic migraine treatment [25]. TENS, PENS, and rTMS are also effective for prophylactic treatment [26]. However, from most literature, the studies are small and many are unblinded [27]. Studies have been inconsistent, rarely reporting full details, which negatively affects meta-analyses. Despite inconsistent results from studies, there are several

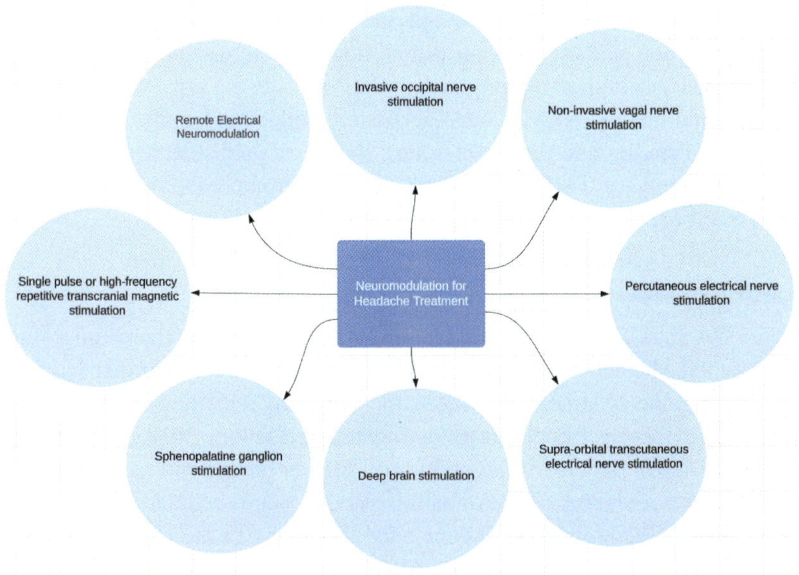

FIGURE 14.1 Neuromodulation techniques for migraine treatment.

Table 14.1 Type of stimulation.

Remote electrical nerve (REN) stimulation inducing conditioned pain modulation	Arm	Acute	Yes	Yes
Invasive electrical stimulation	Great occipital nerve	Prophylactic	No	No
Transcutaneous electrical nerve stimulation (TENS)	Supra-orbital nerve or occipital nerve	Prophylactic	Yes	Yes
High frequency repetitive TMS	Dorsolateral prefrontal cortex and vertex, primary motor cortex (M1)	Prophylactic	Yes	Yes
Percutaneous electric nerve stimulation (PENS)	Shaoyang or Taiyang acupoints	Prophylactic	Yes	Yes
Non-invasive vagus nerve stimulation	Cervical vagus nerve	Acute and prophylactic	Yes	Yes
Single pulse transcranial magnetic stimulation (TMS)	Occipital cortex	Acute	Yes	Yes
Deep brain stimulation	Posterior hypothalamus and ventral tegmentum	Acute and prophylactic	No	No

recent studies published with higher quality. It is recommended in the future for clinical trials to optimize their design and methods. Administration of TMS, subject selection, and outcome measures should be more standardized. Studies should consider multiinstitutional collaboration, using the same equipment, parameters and methodology to most accurately collect and analyze. Therefore, more robust studies and further literature should be published to further determine the benefit of TMS compared to other neuromodulation techniques in terms of efficacy.

Future directions of TMS in headaches

As an FDA and CE approved intervention for acute and prophylactic migraines, Transcranial Magnetic Stimulation has excellent potential for further intervention in migraine treatment. However, the literature regarding TMS varies in quality of evidence. According to a Cochrane review in 2018, there is low-quality evidence for rTMS and short-term effects on chronic pain in general [28]. A 2016 review article evaluating various neuromodulation techniques for headaches (including TMS) concluded that many of the studies published by then were inadequate [26]. The majority of studies were small or did not have an appropriate placebo, which limited robust analysis. A 2019 systematic review evaluating transcranial magnetic and direct current stimulation (TMS/tDCS) for headaches used the GRADE tool to determine the quality of evidence, and the review determined the evidence to range from very low to moderate for outcome measures [1].

One of the major barriers to conducting a strong clinical trial is developing an adequate placebo. The neuromodulation adverse effects may unblind the subject, and the placebo stimulation is difficult to imitate these similar adverse effects. Prior studies have also used sham stimulation electrical currents below the level that would provide clinical efficacy. Lower levels of electrical currents have also not been studied, and so, are not known to be clinically efficacious. Therefore, some studies could have used an active placebo, which would complicate data interpretation.

There is also variation in protocol stimulation parameters, location and duration of treatment, and outcome measures between studies [1]. However, rTMS studies should look into higher frequency use (>5 Hz), subresting motor threshold stimulation (<100%) over the left motor cortex or left dorsolateral prefrontal cortex (DLPFC) because literature is lacking in these method parameters [1]. Many studies also had insufficient blinding of participants or outcome assessors as well as occasional reporting bias. In the future, larger and more robust studies with standardized protocols and better methods of blinding need to be conducted to further analyze the benefit of TMS in headache disorders. If there is more robust evidence, neurostimulation can become a strong potential treatment alongside pharmacological treatment for each headache disorder present.

From the established evidence available, TMS has a role in the treatment of migraines, especially in patients at risk of overuse medication headache or who have migraines refractory to pharmacological treatment. However, in certain patient populations, literature is lacking in safety data. For example, there is no clear safety data from studies on the use of TMS in children or in pregnant women [26]. However, in a 2020 randomized control, pregnant women with major depression disorder had TMS [29], and from this study, TMS in pregnant women is generally safe. In addition, a 2015 postmarket pilot program of the SpringTMS, a sTMS device, three women in the study group completed normal pregnancies per postmarketing data [30]. Although it appears that TMS is relatively safe for pregnant women but further literature should be established to confirm. Therefore, further studies should be published regarding the safety profile of TMS in children and pregnant women.

Another future direction for TMS and other neuromodulation techniques is affordability. Many of these procedures are expensive, so providers and investigators should look into cost reduction in the future. Patients may not be able to benefit from neuromodulation if it is not affordable.

As mentioned previously, future literature can also evaluate the efficacy of TMS versus other neuromodulation techniques for various headache disorders. Many studies were small, not blinded, did not always report full details, and did not have standardization of the methods. It is important to evaluate how effective TMS treatment is compared to other neuromodulation techniques for headache treatment. Therefore, future studies should be larger, more standardized, blinded, and optimized in methodology in order to properly and accurately compare various neuromodulation techniques used for headache intervention.

The majority of the evidence involves the use of TMS for migraine treatment. However, there is a significant opportunity to further investigate TMS' use for headache disorders outside of migraines such as tension-type, cervicogenic, and posttraumatic headaches. Further high-quality randomized-controlled trials with standardized protocols should investigate neuromodulation techniques such as TMS in nonmigrainous headache disorders.

TMS and other neuromodulation techniques have the potential as alternatives to pharmacologic treatment for various headache disorders. Their milder side effect profiles are advantageous in addition to their benefit for those with headache disorders refractory to pharmacologic treatment or those with contraindications to other conventional treatments. Neuromodulation such as TMS is promising for headache treatment, but larger and better-designed trials are needed before TMS is widely used as part of mainstay clinical care.

REFERENCES

[1] J.M. Stilling, O. Monchi, F. Amoozegar, C.T. Debert, Transcranial magnetic and direct current stimulation (TMS/tDCS) for the treatment of headache: a systematic review, Headache 59 (3) (2019) 339–357, https://doi.org/10.1111/head.13479.

[2] M. Hallett, Transcranial magnetic stimulation: a primer, Neuron 55 (2) (2007) 187–199, https://doi.org/10.1016/j.neuron.2007.06.026.

[3] R.B. Lipton, S.H. Pearlman, Transcranial magnetic stimulation in the treatment of migraine, Neurotherapeutics 7 (2) (April 2010) 204–212, https://doi.org/10.1016/j.nurt.2010.03.002.
[4] S.K. Aurora, B.K. Ahmad, K.M. Welch, P. Bhardhwaj, N.M. Ramadan, Transcranial magnetic stimulation confirms hyperexcitability of occipital cortex in migraine, Neurology 50 (4) (1998) 1111–1114, https://doi.org/10.1212/wnl.50.4.1111.
[5] S. Rossi, M. Hallett, P.M. Rossini, A. Pascual-Leone, Safety of TMS Consensus Group, Safety, ethical considerations, and application guidelines for the use of transcranial magnetic stimulation in clinical practice and research, Clin. Neurophysiol. 120 (12) (2009) 2008–2039, https://doi.org/10.1016/j.clinph.2009.08.016.
[6] B.M. Clarke, A.R. Upton, M.V. Kamath, T. Al-Harbi, C.M. Castellanos, Transcranial magnetic stimulation for migraine: clinical effects, J. Headache Pain 7 (5) (2006) 341–346, https://doi.org/10.1007/s10194-006-0329-8.
[7] R.B. Lipton, D.W. Dodick, S.D. Silberstein, et al., Single-pulse transcranial magnetic stimulation for acute treatment of migraine with aura: a randomised, double-blind, parallel-group, sham-controlled trial, Lancet Neurol. 9 (4) (2010) 373–380, https://doi.org/10.1016/S1474-4422(10)70054-5.
[8] P.J. Goadsby, R.B. Lipton, M.D. Ferrari, Migraine–current understanding and treatment, N. Engl. J. Med. 346 (4) (2002) 257–270, https://doi.org/10.1056/NEJMra010917.
[9] F. Brighina, A. Piazza, G. Vitello, et al., rTMS of the prefrontal cortex in the treatment of chronic migraine: a pilot study, J. Neurol. Sci. 227 (1) (2004) 67–71, https://doi.org/10.1016/j.jns.2004.08.008.
[10] M. Teepker, J. Hötzel, N. Timmesfeld, et al., Low-frequency rTMS of the vertex in the prophylactic treatment of migraine, Cephalalgia 30 (2) (2010) 137–144, https://doi.org/10.1111/j.1468-2982.2009.01911.x.
[11] M.E. Keck, T. Welt, M.B. Müller, et al., Repetitive transcranial magnetic stimulation increases the release of dopamine in the mesolimbic and mesostriatal system, Neuropharmacology 43 (1) (2002) 101–109, https://doi.org/10.1016/s0028-3908(02)00069-2.
[12] A.P. Strafella, T. Paus, J. Barrett, A. Dagher, Repetitive transcranial magnetic stimulation of the human prefrontal cortex induces dopamine release in the caudate nucleus, J. Neurosci. 21 (15) (2001) RC157, https://doi.org/10.1523/JNEUROSCI.21-15-j0003.2001.
[13] N. Michael, M. Gösling, M. Reutemann, et al., Metabolic changes after repetitive transcranial magnetic stimulation (rTMS) of the left prefrontal cortex: a sham-controlled proton magnetic resonance spectroscopy (1H MRS) study of healthy brain, Eur. J. Neurosci. 17 (11) (2003) 2462–2468, https://doi.org/10.1046/j.1460-9568.2003.02683.x.
[14] J. Kalita, S. Laskar, S.K. Bhoi, U.K. Misra, Efficacy of single versus three sessions of high rate repetitive transcranial magnetic stimulation in chronic migraine and tension-type headache, J. Neurol. 263 (11) (2016) 2238–2246, https://doi.org/10.1007/s00415-016-8257-2.
[15] L. Koski, T. Kolivakis, C. Yu, J.K. Chen, S. Delaney, A. Ptito, Noninvasive brain stimulation for persistent postconcussion symptoms in mild traumatic brain injury, J. Neurotrauma 32 (1) (2015) 38–44, https://doi.org/10.1089/neu.2014.3449.
[16] A. Leung, V. Metzger-Smith, Y. He, et al., Left dorsolateral prefrontal cortex rTMS in alleviating MTBI related headaches and depressive symptoms, Neuromodulation 21 (4) (2018) 390–401, https://doi.org/10.1111/ner.12615.
[17] H. Hodaj, J.P. Alibeu, J.F. Payen, J.P. Lefaucheur, Treatment of chronic facial pain including cluster headache by repetitive transcranial magnetic stimulation of the motor cortex with maintenance sessions: a naturalistic study, Brain Stimul. 8 (4) (2015) 801–807, https://doi.org/10.1016/j.brs.2015.01.416.
[18] J. Ailani, R.C. Burch, M.S. Robbins, Board of Directors of the American Headache Society, The American headache society consensus statement: update on integrating new migraine treatments into clinical practice, Headache 61 (7) (2021) 1021–1039, https://doi.org/10.1111/head.14153.
[19] S.M. McClintock, I.M. Reti, L.L. Carpenter, et al., Consensus recommendations for the clinical application of repetitive transcranial magnetic stimulation (rTMS) in the

treatment of depression, J. Clin. Psychiatry 79 (1) (2018) 16cs10905, https://doi.org/10.4088/JCP.16cs10905.
[20] R. Taylor, V. Galvez, C. Loo, Transcranial magnetic stimulation (TMS) safety: a practical guide for psychiatrists, Australas. Psychiatr. 26 (2) (2018) 189–192, https://doi.org/10.1177/1039856217748249.
[21] T.H. Hsieh, S.C. Dhamne, J.J. Chen, et al., Minimal heating of aneurysm clips during repetitive transcranial magnetic stimulation, Clin. Neurophysiol. 123 (7) (2012) 1471–1473, https://doi.org/10.1016/j.clinph.2011.10.048.
[22] Z.D. Deng, S.H. Lisanby, A.V. Peterchev, Electric field depth-focality tradeoff in transcranial magnetic stimulation: simulation comparison of 50 coil designs, Brain Stimul. 6 (1) (2013) 1–13, https://doi.org/10.1016/j.brs.2012.02.005.
[23] U. Damar, H. Lee Kaye, N.A. Smith, P.B. Pennell, A. Rotenberg, Safety and tolerability of repetitive transcranial magnetic stimulation during pregnancy: a case report and literature review, J. Clin. Neurophysiol. 37 (2) (2020) 164–169, https://doi.org/10.1097/WNP.0000000000000552.
[24] V. Tiwari, S. Agrawal, Migraine and neuromodulation: a literature review, Cureus 14 (11) (2022) e31223, https://doi.org/10.7759/cureus.31223.
[25] X. Moisset, B. Pereira, D. Ciampi de Andrade, D. Fontaine, M. Lantéri-Minet, J. Mawet, Neuromodulation techniques for acute and preventive migraine treatment: a systematic review and meta-analysis of randomized controlled trials, J. Headache Pain 21 (1) (2020) 142, https://doi.org/10.1186/s10194-020-01204-4.
[26] S. Miller, A.J. Sinclair, B. Davies, et al., Neurostimulation in the treatment of primary headaches, Practical Neurol. 16 (2016) 362–375.
[27] M.M. Klein, R. Treister, T. Raij, et al., Transcranial magnetic stimulation of the brain: guidelines for pain treatment research, Pain 156 (9) (2015) 1601–1614, https://doi.org/10.1097/j.pain.0000000000000210.
[28] N.E. O'Connell, L. Marston, S. Spencer, L.H. DeSouza, B.M. Wand, Non-invasive brain stimulation techniques for chronic pain, Cochrane Database Syst. Rev. 3 (3) (2018) CD008208, https://doi.org/10.1002/14651858.CD008208.pub4.
[29] D.R. Kim, E. Wang, B. McGeehan, et al., Randomized controlled trial of transcranial magnetic stimulation in pregnant women with major depressive disorder, Brain Stimul. 12 (1) (2019) 96–102, https://doi.org/10.1016/j.brs.2018.09.005.
[30] R. Bhola, E. Kinsella, N. Giffin, et al., Single-pulse transcranial magnetic stimulation (sTMS) for the acute treatment of migraine: evaluation of outcome data for the UK post market pilot program, J Headahe Pain 16 (2015).

Chapter | Fifteen

Infusion therapies: Lidocaine, ketamine, magnesium, dihydroergotamine

Anthony J. Batri[1], Melissa M. Sun[2], Danielle N. Nguyen[3], QueenDenise Okeke[4], Spencer Brodsky[5]

[1]*New York-Presbyterian Hospital, the University Hospitals of Columbia and Cornell, Columbia University Irving Medical Center, New York, NY, United States;* [2]*Penn State Health Milton S. Hershey Medical Center, Rowan Virtua School of Medicine, Hershey, PA, United States;* [3]*St. Elizabeth's Medical Center, Boston, MA, United States;* [4]*Philadelphia College of Osteopathic Medicine, Philadelphia, PA, United States;* [5]*Montefiore Medical Center, Albert Einstein College of Medicine, Bronx, NY, United States*

INTRODUCTION

Throughout the years, headache disorders have affected a significant portion of the global population causing a range of symptoms from transient discomfort to debilitating pain and functional impairment. Migraine is one of the most common neurologic disorders, affecting approximately 15% of the global population making it the third most prevalent disorder worldwide. This disorder can be categorized as episodic or chronic based on frequency of headaches in 1 month; at least eight migraine days per month to as many as 15 migraine days per month [1]. Within the migraine class of headaches, about one in 10 patients have the most extreme and more debilitating form, chronic migraine (CM). The global prevalence of CM is approximately between 1% and 2% of the global population [2]. Within this group of patients who have CM, there are individuals who suffer from "refractory migraine" or "intractable migraine". There is currently no universally accepted definition of "refractory migraine" or "intractable migraine", but the consensus agrees that it is a

Interventional Management of Migraines and Other Headache Disorders. https://doi.org/10.1016/B978-0-443-23557-3.00015-2

form of CM that is unresponsive to several classes of commonly used migraine medication [3].

In recent years, infusion therapies utilizing agents such as dihydroergotamine (DHE), lidocaine, ketamine, and magnesium have emerged as promising options for managing severe and chronic headache conditions. This book chapter aims to explore the current landscape of infusion therapies for headache treatment, specifically diving into the use of lidocaine, ketamine, magnesium, and DHE. By delving deeper into the indications of treatment, clinical efficacy, and safety profiles of these infusion therapies, we can gain an improved and comprehensive understanding of their potential to revolutionize headache management.

Before investigating further into more current infusion therapies used, it is important to touch on common treatment methods used including "The Raskin Protocol" developed by Dr. Neil Raskin. In the 1980s, Dr. Raskin first created this protocol that is now well-known and have since had many modified renditions of the original protocol [4]. The protocol describes a systematic and structured approach to discontinuing or tapering medications that may be contributing to the development and perpetuation of chronic daily headaches. The primary focus of this protocol is to address the issue of medication overuse, which occurs when the frequent or excessive use of headache medications leads to a paradoxical increase in headache frequency and severity.

In 1986, Dr. Raskin's experiment involved the study of intravenous dihydroergotamine as a therapy for "intractable migraines". There were 55 patients, of whom 36 were dependent on ergotamine, analgesics, diazepam, or corticosteroids and were given IV DHE and metoclopramide every 8 hours. This group was compared with 54 patients who were age and sex matched to the DHE-treated group, of whom 38 were dependent on the former drugs mentioned [4]. In conclusion, 49 of the 55 DHE-treated patients became headache-free within 48 hours of whom 39 of the 49 had long-term benefits for an average of 16 months. In contrast, only seven diazepam-treated patients because headache-free within three to 6 days, and 31 patients had minimal symptomatic improvements within 10 days. In summary, Dr. Raskin's experiment in 1986 showed potential for repetitive IV DHE in the treatment of "intractable migraines" or "retractable migraines". After this report, the usage of IV DHE has been further explored along with other infusion therapies including lidocaine, ketamine, and magnesium.

Lidocaine (lignocaine) infusion therapy

Lidocaine infusion has emerged as a valuable treatment option for various types of headaches, including migraines and cluster headaches. However, it first was created by Swedish chemists Nils Löfgren and Bengt Lundquist in 1943 as a local anesthetic and antiarrhythmic [5]. Lidocaine was the first in its class of local anesthetics to be classified as an aminoamide as opposed to the other local anesthetics at the time that were aminoesters. Thus, after the development of lidocaine, the creation of other aminoamides came about including bupivacaine, prilocaine, mepivacaine, and ropivacaine [6]. Lidocaine was also the first sodium channel blocker to be identified, specifically by reversibly blocking the voltage-gated sodium channels (VGSC/NaVs) therefore inhibiting the propagation of action potentials at nerve-endings [7]. This effect led to the reversible block of pain signals in the nervous system, providing temporary relief. The mechanism of action of lidocaine includes the blockage of potassium ion channels as well as the regulation of calcium through other ligand-gated ion channels [7]. Since its development, IV lidocaine has been used in treatment for many various chronic pain conditions on top of its subcutaneous administration as a local analgesic effect.

In the mid-2000s, IV lidocaine infusions in the treatment of severe headache disorders gained attention. Hand and Stark performed a retrospective study where they evaluated 19 patients (16 women, three male) who received a total of 27 lidocaine infusions [8]. Out of the 27 infusions, 22 infusions were given to 18 patients with analgesic rebound headaches with a headache resolution in 82% of lidocaine infusions. Subsequently, four patients obtained long-term relief, six patients returned to their original manageable pattern of headaches (two of which however had reoccurrence of chronic daily headaches), four patients were lost to follow-up, and in the last four patients there were no long-term benefits. Of note, during four infusions, seven minor adverse events were recorded.

Stark continued to expand on his work on investigating lidocaine infusion therapies as well as other researchers including Rosen et al. who analyzed the therapeutic effect of IV lidocaine in 68 patients diagnosed with chronic daily headaches or chronic cluster headaches in a retrospective, uncontrolled study through reviewing medical records between 2003 and 2005. The studies mentioned investigated the use of lidocaine infusion therapy at various titrations ranging from 2 mg/minute over a period of 7–10 days to serum concentration of less than 5 μg/mL over a period of 2–15 days [9]. However, the studies all showed positive efficacy of IV lidocaine in the treatment of headache disorders in varying degrees and is an area of interest and future studies will be beneficial to elucidate more of the therapeutic effects.

Ketamine infusion

Ketamine was initially developed as an anesthetic agent and has shown promise in headache management due to its N-methyl-D-aspartate (NMDA) receptor antagonism and modulation of pain pathways. Additionally, ketamine has received attention due its dissociative effects that is caused by the stimulation of the frontal cortex and simultaneous suppression of the thalamus [10]. Over the past few decades, studies have increasingly focused on the role of ketamine infusion in treating refractory migraines and chronic daily headaches, revealing encouraging outcomes in terms of pain reduction and improved functional abilities. In Schwenk et al.'s retrospective study in 2018, 61 patients admitted over 3 years whom received 5 days of intravenous therapy including continuous ketamine of a mean maximum ketamine rate of 0.76 mg/kg per hour. The results of this study showed that 77% of the patients were immediate responders showing an association with short-term analgesia in many patients with chronic refractory headaches [11]. While ketamine's analgesic effect on chronic migraines has promise, there is still a need for more controlled longitudinal studies on the long-term and adverse effects of ketamine on headaches.

Magnesium infusion

Magnesium, an essential mineral involved in numerous physiological processes, has gained attention as a potential treatment for migraines, especially in patients with magnesium deficiencies. Many studies have shown magnesium to be a well-tolerated, cost-effective, and safe option for pain-relief in headache disorders [12]. However, the studies looking particularly at intravenous magnesium in acute migraine treatment have been inconsistent. In 2001, a single-blind placebo-controlled, randomized trial of 30 patients with migraines were randomly given placebo or 1 mg magnesium sulfate. The treatment response of magnesium sulfate was superior to placebo in both analgesic effect as well as response rate [13]. Another study in 2002 that also studied the effect of 1 mg of magnesium sulfate in migraine patients had results that corroborated magnesium's positive effect on pain improvement [13]. Based on several papers and studies done throughout the years, the role of magnesium is well-documented, both in oral and intravenous forms. Some of these studies touched upon the association between migraines and magnesium deficiency as well. However, clinical judgment should be used when deciding when and whom to use magnesium treatments.

Dihydroergotamine infusion

Dihydroergotamine (DHE) was first created in the 20th century as a derivative of ergotamine. It has been utilized for decades in the management of migraines and cluster headaches. DHE, compared with ergotamine, has a greater effect in its antagonist effect on alpha-adrenergic activity. In the 1920s, ergotamine was first used in obstetrics and gynecology as a uterotonic [14]. Twenty years later, DHE was developed and had less arterial vasoconstriction, lower dopaminergic agonism, and decreased emetic effect compared with ergotamine. For many years since, both ergotamine and DHE were considered the only migraine-specific treatment until triptans came onto the scene.

Infusion therapies, such as lidocaine, ketamine, magnesium, and DHE have proven to be important tools in the management and treatment of various headache disorders. We discuss the indications, techniques, efficacy, and safety of these infusions in this chapter.

INDICATIONS FOR INFUSIONS

Infusion or intravenous therapy involves administering medications, fluids, electrolytes, and/or nutrients into the body through a vein. Though not commonly considered first line treatment for headaches, infusion therapy may be utilized for severe or refractory headaches that have failed various treatment modalities. Infusion therapy may be used to treat a variety of headache types, including migraines, cluster headaches, tension headaches, and status migrainosus.

Migraines

The pathophysiology of migraines is intricate and characterized by heterogeneity, leading to a diverse range of presentations and variable responses to different treatment approaches. One emerging modality is the use of intravenous lidocaine infusion. Schwenk et al. conducted a retrospective analysis on refractory chronic migraine patients who received lidocaine infusions, involving a total of 609 participants [15]. The primary outcome measured the change in headache pain from baseline to hospital discharge. The results showed a significant median pain reduction from a baseline of 7 to 1 ($P < .001$) at the end of hospitalization. The study concluded that lidocaine infusions could provide short-term and medium-term pain relief in refractory chronic migraine. However, side effects of lidocaine infusion, such as transient bradycardia/junctional heart rhythm, nausea, hallucinations, blurry vision, and insomnia, were reported [15].

In another investigation by Schwenk et al., patients with refractory chronic migraine were administered a continuous ketamine infusion over a span of 5 days [16]. At baseline, the mean pain intensity was 7.4 ± 1.4, which decreased to 3.7 ± 2.3 ($P \leq .05$) by day five. Side effects of ketamine infusion included nausea, vomiting, vivid dreams, hallucinations, nightmares, and blurry vision.

Dihydroergotamine infusion is another commonly utilized treatment for migraines. In a study by Charles et al., 31 patients received continuous dihydroergotamine infusion over 3 days [17]. The primary outcome assessed the change in pain intensity on an 11-point numeric scale. At the end of the 3-day period, there was an average decrease of 63.4% in pain intensity, with approximately one third of patients experiencing no headache after day three. Long-term follow-up showed an approximately 86% decrease in headache frequency. It is important to note that although this study demonstrated the efficacy of outpatient continuous dihydroergotamine infusion, caution must be exercised as it can cause vasoconstriction, potentially necessitating inpatient hospitalization for closer monitoring in individuals with advanced age or cardiovascular comorbidities [18].

Cluster headaches

Cluster headaches, a highly debilitating form of primary headache disorder, often require prompt and effective treatment. Intravenous infusion therapy has emerged as a potential therapeutic approach to alleviate the intense pain and associated symptoms experienced by individuals suffering from cluster headaches. A study performed by Magnoux et al. studied intravenous dihydroergotamine treatment of refractory cluster headache in 70 patients [19]. In the case of episodic cluster headache, pain was completely resolved at 1 month in 63% of cases, partially resolved in 15% of cases, and failed to resolve in 24% of cases whereas in the case of chronic cluster headache, pain was completely resolved at 1 month in 46% of cases, partially resolved in 11% of cases, and failed to resolve in 43% of cases. Of note, intravenous dihydroergotamine caused chest pain, palpitations, chest tightness, legs cramps, nausea, and diarrhea in a couple patients involved in this study.

More conventional approaches, such as intravenous dihydroergotamine, have shown varying degrees of success in alleviating symptoms in patients with refractory chronic cluster headaches. In light of the limited effectiveness of conventional treatments, there is a growing need to explore innovative therapeutic strategies for refractory chronic cluster headache. Xavier et al. utilized one such novel approach by combining a ketamine infusion with magnesium sulfate in a study with 17 patients. Out of 17 patients, 13 patients, or 76%, responded well to treatment. In these patients, the total amount of daily attacks decreased from 4.3 ± 2.4 before treatment to 1.3 ± 1.0 after treatment (difference: -3.1 (95% CI: -4.5 to -1.6), $P < 0.001$) [20]. Side effects in these patients included transient and mild sedation. Despite these side effects, the combination of ketamine and magnesium sulfate infusion offers a promising avenue for addressing this debilitating condition, with a notable reduction in attack frequency and tolerable side effects. Further research and clinical trials are warranted to validate these results and explore the full potential of this innovative treatment approach.

Tension headaches

Tension headaches are a common type of headache typically characterized by a dull, aching bilateral pain, often diffusely extending over the top of the head. Though not usually quite as severe as migraines, they can be uncomfortable and affect people's abilities to perform their daily activities. Lidocaine infusion is a treatment modality that has been explored for tension headaches, particularly in cases of refractory tension headaches that have not responded well to other therapies or to decrease the used of nonsteroidal antiinflammatory drugs in the ED. In a study by Akbas et al., lidocaine infusion versus dexketoprofen infusion administered for 15 minutes was compared in the treatment of episodic tension type headache [21]. One hundred and twenty patients were randomized into the lidocaine and dexketoprofen groups. Median age of the study groups was 43 and the median duration of pain was 120 minutes in the lidocaine infusion group versus 160 minutes in the dexketoprofen infusion group. The change in pain score in a one, two, three, four, and 5 week period were statistically significantly higher in the lidocaine group than in the dexketoprofen group. The adverse events experienced after these transfusions were transient and resolved. Some patients in the lidocaine group experienced skin rash and burning in the arm upon transfusion while some patients in the deketoprofen group experiences itching and burning in the arm upon transfusion. There was no statistically significant difference in experiences adverse effects between the two study groups within the 1 week follow up period. This study suggests that lidocaine infusions may be associated with significantly greater decreases in episodic tension type headache pain than systemic nonsteroidal antiinflammatory medications. Though lidocaine infusions are not standard or first-line treatment for episodic tension-type headaches, they show promise in treating episodic tension type headaches. More research is required to establish its efficacy, safety, and appropriate patient selection.

Status migrainosus

Status migrainosus refers to a severe and prolonged migraine attack that lasts for more than 72 hours. It is considered a medical emergency due to its debilitating nature and resistance to conventional treatments. In such cases, alternative therapeutic options are often sought to provide prompt relief. There has been growing interest in utilizing intravenous magnesium as abortive therapy for migraines as it has been suggested that magnesium deficiency may be one of the factors that contributes to the underlying mechanisms of migraine pathophysiology. In a retrospective study by Xu et al., 234 status migrainosus patients were treated with IV magnesium for abortive therapy [22]. After receiving the magnesium infusion, pain score decreased from 5.46 ± 2.39 to 3.56 ± 2.75 ($P < .001$). Of the 234 patients, 54% had clinically significant decreases in pain, or a decrease in pain $\geq$30%. Of note, patients with a lower baseline pain scale rating tended to respond better to treatment than patients with a higher baseline pain scale since patients who achieved pain reduction $\geq$30% had significantly lower pretreatment pain scores ($P = .018$). In this study, IV magnesium therapy showed promising results in providing clinically significant pain relief for a subgroup of patients with status migrainosus, potentially suggesting that IV magnesium could be a cost-effective parenteral therapy for status migrainosus in patients who have lower initial pain intensity at baseline. Although IV magnesium has shown promise as an abortive therapy for status migrainosus, it is important to note that individual responses to treatment may vary. Additionally, the use of IV magnesium should be carefully considered in patients with certain medical conditions, such as impaired renal function or heart disease, as magnesium levels need to be monitored to prevent toxicity.

OVERVIEW OF INFUSION TECHNIQUES

Infusion therapies using lidocaine, ketamine, magnesium, and dihydroergotamine (DHE) are gaining popularity as a treatment for chronic headaches. These therapies are currently being researched and applied in clinical settings. However, obtaining consistent and successful therapeutic outcomes depends on establishing standardized protocols for administration, including dosage, infusion rate, and treatment duration, to achieve the desired therapeutic effect. In this section, we will review several studies and their administration protocols, such as intermittent, repetitive, or continuous infusions, to determine the most effective techniques for treating headaches.

Techniques for lidocaine infusion

The administration of intravenous (IV) lidocaine infusion has been found to be an effective treatment method for chronic headaches, including migraine and cluster headaches. Williams and Stark (2003) were the first to describe the use of IV lidocaine as a continuous infusion for managing chronic headaches from medication overuse [23]. In their retrospective cohort study, 71 patients with medication overuse headaches (ranging from migraine, chronic tension-type, and daily persistent headache) were treated with IV lidocaine at a continuous infusion rate of 2 mg/min. Patients were monitored on 24-hour bedside EKG, and in cases where lidocaine infusion was not sufficient in controlling breakthrough headaches, patients were prescribed NSAIDs such as Diclofenac 50 mg or Naproxen 500 mg, with a maximum of three doses per day. Results showed that 90% of patients experienced absence or improvement in their headaches by the end of the 8.7-day treatment, 97% of patients reported successful withdrawal from the implicated medication, and 70% had absence or improvement of headache at 6 months [13,24]. Results concluded that intravenous lidocaine infusion can effectively reduce the severity and frequency of chronic daily headaches, as well as the medication overuse associated with them.

Rosen et al. expanded on the work of Williams and Stark. They retrospectively studied a similar patient population (n = 68) over a period of 2.5 years [23,25]. Their protocol adjusted the lidocaine infusion rate, time, and treatment duration. The infusion began at 1 mg/min for 4 h, then increased to 2 mg/min. The lidocaine dose was then adjusted based on the patient's response and tolerance, after reviewing their serum lidocaine levels, with a maximum lidocaine level of 5 mg/mL[3]. Study results reported a mean decrease of four points on a 10-point pain scale after multiday lidocaine infusion. No reports on outcomes beyond hospital discharge were reported, so it is unknown if patients who acutely responded in the hospital had sustained progress [13,25].

In light of the high cost of inpatient hospitalization, researchers agree that shorter treatment lengths, hospital stays, and outpatient infusions would allow IV lidocaine infusions to be more feasible in treating more people with chronic headaches [13]. While outpatient use of IV lidocaine infusions has not been well studied, two studies investigated the efficacy of IV lidocaine infusion in treating migraine attacks in the emergency department [13,26–28]. Bell et al. used IV lidocaine 50 mg at 20-minute intervals to a maximum total dose of 150 mg and found that only 29% of patients reported relief at 2 hours [13,28]. In 2021, Gur et al. investigated the efficacy of 1-hour infusions of lidocaine in treating migraine attacks in emergency departments [26,27]. Patients received a 1.5 mg/kg lidocaine bolus and a 1 mg/kg lidocaine infusion (first 30 minutes), followed by a 0.5 mg/kg intravenous infusion for the remaining 30 minutes [27]. The results of the study showed that using IV lidocaine reduced migraine pain scores and led to fewer revisits to the ER as compared with standard nonsteroidal antiinflammatory drugs (NSAIDs) treatments [26]. The benefits of repeated outpatient IV lidocaine infusions need further investigation and continue to be a topic for future studies.

Techniques for ketamine infusion

IV Ketamine infusion in treating various pain conditions and mood disorders has been well established [29]. Recently, case series have focused on the role of IV ketamine infusion in treating headaches, such as refractory chronic migraine. Results found that with carefully monitored settings, the inpatient administration of IV ketamine can successfully treat refractory chronic migraine [9]. Krusz et al. treated 30 patients with 0.4 mg/kg IV ketamine infusion over 90 minutes, with a second infusion given if there were no side effects observed. The average ketamine dose was 71 mg and the average infusion time was 142 minutes. Treatment demonstrated improvement in pain scores, from 6.61/10 to 3.4/10 ($P < .001$) after treatment, with few side effects recorded [30]. The retrospective study by Shcwenk et al. treated 61 patients over 5 days with a continuous infusion of ketamine. The mean maximum ketamine rate of 65.2 ± 2.8 mg/hour (0.76 mg/kg per hour). Results established that IV ketamine infusion was associated with immediate short-term pain relief in 77% of refractory headaches and patients reported minimum and tolerable adverse effects [31]. A smaller retrospective case series by Lauritsen et al. administered IV ketamine following their standard protocol. Dosing started at 0.1 mg/kg/hour and increased by 0.1 mg/kg/hour every 3–4 hours, depending on the patient's tolerance until the target pain score of 3/10 was reached and maintained for 8 hours. The average ketamine infusion rate during this target period was 0.34 mg/kg/hour (range 0.12–0.42 mg/kg/hour) [32]. According to the results, all patients achieved the target pain level of three or less for 8 hours with no significant side effects reported. However, the authors did report poor patient follow-up posttreatment therefore, the efficacy of treatment long term could not be established.

These studies reveal promising outcomes of IV ketamine infusion in reducing migraine pain. However, the literature shows variation in dosing strategies, therefore, care must be taken to adjust doses and infusion rates based on patient characteristics as well as treatment response [29].

Techniques for magnesium infusion

The role of IV magnesium infusion in treating headaches such as migraine has mostly been described as a well-tolerated, safe, and inexpensive therapy [33]. However, protocols for the most effective dosage, duration, and infusion rates remain limited and inconclusive. In an outpatient retrospective study, Xu et al. treated 234 patients with migraine headaches. In their protocol, these patients received 2g of magnesium sulfate (diluted with 50–100 cc of normal saline) infused over 1–2 hours. The authors reported that the overall patient pain score decreased from 5.46 ± 2.39 to 3.56 ± 2.75 ($P < .001$) after magnesium infusion [22]. Demirkaya et al. conducted a study with 30 patients using different magnesium regimens. They demonstrated that 1g of IV magnesium sulfate infused over 15 minutes led to therapeutic pain relief of severe migraine [13]. Despite the differences in sample size, both studies reported IV magnesium sulfate to be an effective, safe, and well-tolerated treatment for severe migraine headaches.

In contrast, the meta-analysis by Choi and Parmar evaluated 295 patients with acute migraine. They reported that 1g or 2g of IV magnesium sulfate infused over 30 minutes did not provide adequate pain relief yet increased patient risk for side effects when compared with the placebo/control group [34].

Due to the cost-effective and low-risk profile of magnesium, clinicians are exploring the application and potential benefits of IV magnesium in the management of migraine headaches [22,33]. It has been noted that studies on the effectiveness of IV magnesium in treating migraines and chronic headaches have produced varying results due to differing protocols. To better understand the ideal dose, duration, and infusion rates, it would be beneficial to conduct large-scale studies with standardization in magnesium infusion protocols and dosages.

Techniques for dihydroergotamine (DHE) infusion

The efficacy of IV Dihydroergotamine (DHE) infusion in the management of chronic headache has been well-documented in both inpatient and outpatient clinical settings. Despite this, DHE is a less utilized medication perhaps due to its unfamiliarity amongst clinicians and the lack of standardization in DHE infusion protocols [35,36].

The protocol outlined by Raskin et al. involved the administration of repetitive IV DHE therapy to 55 patients who suffered from chronic headaches. This protocol was carried out over a period of 2 days while the patients were hospitalized. Patients were pretreated with 10 mg IV metoclopramide to prevent nausea, and the average dose of IV DHE was 0.7 mg (with a range of 0.3–1.0 mg), repeated every 8 hours. Minimal adverse effects were reported, and results showed that 49 out of 55 DHE-treated patients became headache-free within 48 hours. Additionally, 39 patients reported a sustained reduction in headache pain and frequency, and 17 patients remained headache-free at 16-month follow-up [4].

Ford later expanded on the work of Raskin et al. He published a 5-year study with 171 patients with refractory headaches treated in the hospital for 7 days. The protocol for repetitive IV DHE followed the published Raskin protocol. The protocol for continuous IV DHE gave patients 3 mg of IV DHE at a continuous rate of 42 mL/hour, with the rate of infusion adjusted (to a range of 21–30 mL/hour) based on the patient's tolerance to side effects such as nausea. Results were comparable between the two patient groups. For patients treated with continuous IV DHE, 92.5% became headache-free, 64.5% within 3 days.

For patients treated with repetitive IV DHE, 86.5% became headache-free, 66.5% within 3 days [37]. The average hospital stay was 4 days, side effects were minimal and tolerable for most patients, and long-term follow-up beyond 7 days was not recorded [17,37]. Ford's study demonstrated that continuous IV DHE is an effective and safe method of treatment that produces results similar to repetitive IV DHE.

Furthermore, in 2010, Charles and Dohln conducted the first outpatient home-based study that treated 31 patients with chronic migraine and medication overuse headaches with continuous DHE infusions for 3 days [17]. According to their protocol, patients were pretreated with 10 mg IV metoclopramide before receiving a continuous infusion of 3 mg DHE at a rate of 42 mL/h on days 1 and 2. On day 3, DHE infusion was reduced to 1.5 mg at a rate of 21 mL/h. By the end of day 3, patients reported an average decrease of 63.4% in headache pain intensity, with one-third of patients reporting no headaches at all. Long-term follow-up showed an average 86% reduction in headache frequency, and patients with medication overuse headaches no longer consumed the offending medication [17]. The authors concluded that continuous IV DHE is a safe and effective therapy for chronic headaches in an outpatient setting, and no serious adverse effects were reported.

Intravenous DHE infusion therapy is both effective and safe in treating chronic headache conditions in hospital and outpatient settings. Evidence-based data has demonstrated that treatment outcomes may be achieved using either repetitive or continuous IV DHE infusions, with some experts claiming that continuous IV DHE infusion is a more convenient method that confers slightly better results [17,37]. Nevertheless, successful therapeutic outcomes depend heavily on determining the appropriate dosage, infusion rate, and duration of IV DHE treatment needed to be effective without producing side effects.

EFFICACY AND SAFETY OF INFUSIONS

Overview

Infusion therapy often serves as an effective option when treating headache disorders that necessitate emergency department evaluation, hospital admission, and those resistant to first-line treatments. Intravenous lidocaine, ketamine, magnesium sulfate, and dihydroergotamine (DHE) have all been proposed as options for these challenging clinical scenarios. While serious adverse events are rare, some carry significant risks that clinicians should carefully consider when starting. Furthermore, nonserious but undesirable side effects are very common with these infusion therapies, which limit their wide-spread use.

Lidocaine

Lidocaine infusion has been studied as an effective treatment option for many headache disorders including migraine, medication overuse headache, cluster headache, and postprocedural headache [23–25,38]. Additional headache disorders where IV lidocaine has been studied include trigeminal neuralgia and

SUNCT syndrome [23,39,40]. A retrospective cohort study by Ray et al. [27] examined the effects of infusion therapy in patients admitted to a tertiary care hospital with a headache disorder. Of the 45 patients in the migraine cohort, more than half (N = 23, 51.1%) experienced at least a 50% reduction in their pain as per the Visual Analogue Scale (VAS) after the lidocaine infusion. For these patients, there was an average reduction of 53% (standard deviation of 39.3) in their VAS pain scores. Moreover, 31.1% (N = 14) reported being completely free from pain, on average, 6.2 days posttreatment (SD = 2.4 days).

Adverse events are common in patients treated with intravenous lidocaine. Through its inhibitory action at the level of the central nervous system, IV lidocaine frequently causes neuropsychiatric reactions, including perioral paresthesia, feelings of euphoria, light headedness, agitation, and tremors [41]. These side effects are typically dose-dependent and resolve after stopping the infusion [41]. Notably, lidocaine has a long-history as an antiarrhythmic, and therefore monitoring cardiac function is crucial in patients receiving IV lidocaine [41]. In a literature review representing approximately 6000 individuals, Gil-Gouveia R et al. [41] report the rate of severe adverse events, including seizures, coma, severe heart block, and cardiac arrest, to be between 0.3% and 33.3%.

Ketamine

In Ray et al.'s retrospective cohort study, ketamine infusion was found, like lidocaine, to be effective in managing headache in the inpatient setting [27]. In the migraine cohort, the authors reported that patients receiving intravenous ketamine had an average reduction in VAS pain scores of 31.6% (SD 41.3), with 34.4% (11/32) achieving 50% or greater reduction in VAS scores over an mean of 3.4 days (SD 0.5).

Ketamine, known for its dissociative anesthetic properties, is frequently used in rapid sequence intubation and procedural sedation, making it particularly suitable for procedures that necessitate conscious sedation. These characteristics, however, often lead to psychomimetic side effects such as hallucinations and vivid dreams [31]. Further adverse effects can include diplopia, nystagmus, nausea, and hypertension [31]. Various studies suggest that the alpha-2 agonist clonidine can be effective in mitigating both the psychotropic and sympathomimetic effects of ketamine [31,42].

Magnesium

Magnesium deficiency is thought to play an important role in the pathogenesis of migraine [43]. Chiu et al.'s 2016 meta-analysis investigated the efficacy of intravenous magnesium sulfate in treating acute migraines, identifying 11 studies (948 participants). The researchers reported that magnesium sulfate infusions significantly ameliorated acute migraines. In the studies assessed, reductions in acute migraine symptoms occurred at various intervals, including within 15–45 minutes (OR = 0.23), 120 minutes (OR = 0.20), and 24 hours

(0.25) following the initial infusion. Generally, intravenous magnesium sulfate is well tolerated, though some commonly reported side effects include flushing and a burning sensation in the face and neck [44,45].

Dihydroergotamine (DHE)

In general, triptans have surpassed ergot alkaloids as the drug-of-choice for initial migraine management as these agents have enhanced receptor selectivity and typically greater efficacy [44]. However, there is an important role for ergot alkaloids, such as the serotonin agonist dihydroergotamine (DHE), when treating headaches refractory to first-line therapy [45]. Intravenous DHE has been shown to be particularly efficacious in achieving pain-free status in patients with chronic migraine and cluster headache [46]. Additionally, in a study by Pringsheim et al. [47], 47% of patients (n = 62) with medication overuse headache had complete resolution of their symptoms during their hospitalization after starting a standardized DHE infusion protocol. Notably, the authors called study participants 3-month after discharge, and found 46% (n = 61) had a reduction in their headaches by 50% or more. In a study of 163 patients with refractory primary headaches, Nagy et al. [46] found IV dihydroergotamine to be generally well tolerated, with the most common adverse effect being nausea (n = 94). Managing this adverse reaction is crucial, given that in this research, 30 participants reported severe nausea, leading to the cessation of DHE treatment in six instances. In several studies, antiemetics such as metoclopramide are included as part of the DHE protocol, which has helped reduce the risk of this adverse effect [45].

REFERENCES

[1] Headache classification committee of the international headache society (IHS) the international classification of headache disorders, 3rd edition, Cephalalgia 38 (1) (2018) 1−211, https://doi.org/10.1177/0333102417738202.

[2] J. Natoli, A. Manack, B. Dean, et al., Global prevalence of chronic migraine: a systematic review, Cephalalgia 30 (5) (2010) 599−609, https://doi.org/10.1111/j.1468-2982.2009.01941.x.

[3] L. D'Antona, M. Matharu, Identifying and managing refractory migraine: barriers and opportunities? J. Headache Pain 20 (1) (2019) 89, https://doi.org/10.1186/s10194-019-1040-x.

[4] N.H. Raskin, Repetitive intravenous dihydroergotamine as therapy for intractable migraine, Neurology 36 (7) (1986), https://doi.org/10.1212/WNL.36.7.995, 995-995.

[5] T. Gordh, Xylocain, a new local analgesic, Anaesthesia 4 (1) (1949) 4−9, https://doi.org/10.1111/j.1365-2044.1949.tb05802.x.

[6] T. Gordh, T.E. Gordh, K. Lindqvist, D.S. Warner, Lidocaine: the origin of a modern local anesthetic, Anesthesiology 113 (6) (2010) 1433−1437, https://doi.org/10.1097/ALN.0b013e3181fcef48.

[7] X. Yang, X. Wei, Y. Mu, Q. Li, J. Liu, A review of the mechanism of the central analgesic effect of lidocaine, Medicine 99 (17) (2020) e19898, https://doi.org/10.1097/MD.0000000000019898.

[8] P.J. Hand, R.J. Stark, Intravenous lignocaine infusions for severe chronic daily headache, Med. J. Aust. 172 (4) (2000) 157–159, https://doi.org/10.5694/j.1326-5377.2000.tb125538.x.
[9] J.J. Mojica, E.S. Schwenk, C. Lauritsen, S.J. Nahas, Beyond the Raskin protocol: ketamine, lidocaine, and other therapies for refractory chronic migraine, Curr. Pain Headache Rep. 25 (12) (2021) 77, https://doi.org/10.1007/s11916-021-00992-x.
[10] G. Corssen, M. Miyasaka, E.F. Domino, Changing concepts in pain control during surgery: dissociative anesthesia with CI-581. A progress report, Anesth. Analg. 47 (6) (1968) 746–759.
[11] E.S. Schwenk, A.C. Dayan, A. Rangavajjula, et al., Ketamine for refractory headache, Reg. Anesth. Pain Med. 1 (June 2018), https://doi.org/10.1097/AAP.0000000000000827.
[12] L.A. Yablon, A. Mauskop, Magnesium in Headache, 2011.
[13] S. Demirkaya, O. Vural, B. Dora, M.A. Topçuoğlu, Efficacy of intravenous magnesium sulfate in the treatment of acute migraine attacks, Headache 41 (2) (2001) 171–177, https://doi.org/10.1046/j.1526-4610.2001.111006171.x.
[14] R. Shafqat, Y. Flores-Montanez, V. Delbono, S.J. Nahas, Updated evaluation of IV dihydroergotamine (DHE) for refractory migraine: patient selection and special considerations<</p>, J. Pain Res. 13 (2020) 859–864, https://doi.org/10.2147/JPR.S203650.
[15] E.S. Schwenk, A. Walter, M.C. Torjman, S. Mukhtar, H.T. Patel, B. Nardone, G. Sun, B. Thota, C.G. Lauritsen, S.D. Silberstein, Lidocaine infusions for refractory chronic migraine: a retrospective analysis, Reg. Anesth. Pain Med. 47 (7) (July 2022) 408–413, https://doi.org/10.1136/rapm-2021-103180. PMID: 35609890.
[16] E.S. Schwenk, M.C. Torjman, R. Moaddel, J. Lovett, D. Katz, W. Denk, C. Lauritsen, S.D. Silberstein, I.W. Wainer, Ketamine for refractory chronic migraine: an observational pilot study and metabolite analysis, J. Clin. Pharmacol. 61 (11) (November 2021) 1421–1429, https://doi.org/10.1002/jcph.1920. Epub 2021 Jul 9. PMID: 34125442; PMCID: PMC8769496.
[17] J.A. Charles, P. von Dohln, Outpatient home-based continuous intravenous dihydroergotamine therapy for intractable migraine, Headache 50 (5) (May 2010) 852–860, https://doi.org/10.1111/j.1526-4610.2010.01622.x. Epub 2010 Jan 28. PMID: 20132337.
[18] C. Lauritsen, M. Myers, M. Hopkins, S. Silberstein, Safety and efficacy of repetitive dihydroergotamine infusion for the acute treatment of refractory chronic migraine in hospitalized patients with vascular risk factors (4366), Neurology 96 (15 Suppl. ment) (2021) 4366.
[19] E. Magnoux, G. Zlotnik, Outpatient intravenous dihydroergotamine for refractory cluster headache, Headache 44 (3) (March 2004) 249–255, https://doi.org/10.1111/j.1526-4610.2004.04055.x. PMID: 15012663.
[20] X. Moisset, P. Giraud, E. Meunier, S. Condé, M. Périé, P. Picard, B. Pereira, D. Ciampi de Andrade, P. Clavelou, R. Dallel, Ketamine-magnesium for refractory chronic cluster headache: a case series, Headache 60 (10) (November 2020) 2537–2543, https://doi.org/10.1111/head.14005. Epub 2020 Oct 31. PMID: 33128280.
[21] I. Akbas, A.O. Kocak, S.T. Akgol Gur, E. Oral Ahiskalioglu, S. Dogruyol, T. Dolanbay, M. Demir, Z. Cakir, Lidocaine versus dexketoprofen in treatment of tension-type headache: a double-blind randomized controlled trial, Am. J. Emerg. Med. 41 (March 2021) 125–129, https://doi.org/10.1016/j.ajem.2020.12.057. Epub 2021 Jan 7. PMID: 33423013.
[22] F. Xu, A. Arakelyan, A. Spitzberg, L. Green, P.H. Cesar, A. Csere, O. Nworie, S. Sahai-Srivastava, Experiences of an outpatient infusion center with intravenous magnesium therapy for status migrainosus, Clin. Neurol. Neurosurg. 178 (March 2019) 31–35, https://doi.org/10.1016/j.clineuro.2019.01.007. Epub 2019 Jan 21. PMID: 30685601.
[23] T. Berk, S.D. Silberstein, The use and method of action of intravenous lidocaine and its metabolite in headache disorders, Headache 58 (5) (May 2018) 783–789, https://doi.org/10.1111/head.13298. Epub 2018 Mar 14. PMID: 29536530.
[24] D.R. Williams, R.J. Stark, Intravenous lignocaine (lidocaine) infusion for the treatment of chronic daily headache with substantial medication overuse, Cephalalgia 23 (10)

(December 2003) 963–971, https://doi.org/10.1046/j.1468-2982.2003.00623.x. PMID: 14984229.

[25] N. Rosen, M. Marmura, M. Abbas, S. Silberstein, Intravenous lidocaine in the treatment of refractory headache: a retrospective case series, Headache 49 (2) (February 2009) 286–291, https://doi.org/10.1111/j.1526-4610.2008.01281.x. PMID: 19222600.

[26] S.T.A. Gur, E.O. Ahiskalioglu, M.E. Aydin, A.O. Kocak, P. Aydin, A. Ahiskalioglu, Intravenous lidocaine vs. NSAIDs for migraine attack in the ED: a prospective, randomized, double-blind study, Eur. J. Clin. Pharmacol. 78 (2022) 27–33, https://doi.org/10.1007/s00228-021-03219-5.

[27] J.C. Ray, S. Cheng, K. Tsan, H. Hussain, R.J. Stark, M.S. Matharu, E. Hutton, Intravenous lidocaine and ketamine infusions for headache disorders: a retrospective cohort study, Front. Neurol. 13 (March 9, 2022) 842082, https://doi.org/10.3389/fneur.2022.842082. PMID: 35356451; PMCID: PMC8959588.

[28] R. Bell, D. Montoya, A. Shuaib, M.A. Lee, A comparative trial of three agents in the treatment of acute migraine headache, Ann. Emerg. Med. 19 (10) (October 1990) 1079–1082, https://doi.org/10.1016/s0196-0644(05)81507-0. PMID: 2221511.

[29] M. Fischer, A. Abd-Elsayed, Ketamine infusion therapy, in: A. Abd-Elsayed (Ed.), Infusion Therapy, Springer, Cham, 2019, https://doi.org/10.1007/978-3-030-17478-1_2.

[30] J. Krusz, J. Cagle, S. Hall, Efficacy of IV ketamine in treating refractory migraines in the clinic, J. Pain 9 (4) (2008) 30.

[31] E.S. Schwenk, A.C. Dayan, A. Rangavajjula, M.C. Torjman, M.G. Hernandez, C.G. Lauritsen, S.D. Silberstein, W. Young, E.R. Viscusi, Ketamine for refractory headache: a retrospective analysis, Reg. Anesth. Pain Med. 43 (8) (November 2018) 875–879, https://doi.org/10.1097/AAP.0000000000000827. PMID: 29923953.

[32] C. Lauritsen, S. Mazuera, R.B. Lipton, S. Ashina, Intravenous ketamine for subacute treatment of refractory chronic migraine: a case series, J. Headache Pain 17 (1) (December 2016) 106, https://doi.org/10.1186/s10194-016-0700-3. Epub 2016 Nov 22. PMID: 27878523; PMCID: PMC5120050.

[33] A.J. Wendahl, A.L. Weinstein, Magnesium infusion therapy, in: A. Abd-Elsayed (Ed.), Infusion Therapy, Springer, Cham, 2019, https://doi.org/10.1007/978-3-030-17478-1_4.

[34] H. Choi, N. Parmar, The use of intravenous magnesium sulphate for acute migraine: meta-analysis of randomized controlled trials, Eur. J. Emerg. Med. 21 (1) (February 2014) 2–9, https://doi.org/10.1097/MEJ.0b013e3283646e1b. PMID: 23921817.

[35] J. Karri, A. Abd-Elsayed, Dihydroergotamine infusion therapy, in: A. Abd-Elsayed (Ed.), Infusion Therapy, Springer, Cham, 2019, https://doi.org/10.1007/978-3-030-17478-1_7.

[36] S.D. Silberstein, US Headache Consortium, Practice parameter: evidence-based guidelines for migraine headache (an evidence-based review) Report of the Quality Standards Subcommittee of the American Academy of Neurology, Neurology 55 (6) (2000) 754–762.

[37] R.G. Ford, K.T. Ford, Continuous intravenous dihydroergotamine in the treatment of intractable headache, Headache 37 (3) (March 1997) 129–136, https://doi.org/10.1046/j.1526-4610.1997.3703129.x. PMID: 9100396.

[38] D. Schere, S.D. Silberstein, Intravenous lidocaine infusion for the treatment of post-acoustic neuroma resection headache: a case report, Headache 49 (2009) 302–303, https://doi.org/10.1111/j.1526-4610.2008.01145.x.

[39] S.J. Scrivani, A. Chaudry, R.J. Maciewicz, D.A. Keith, Chronic neurogenic facial pain: lack of response to intravenous phentolamine, J. Orofac. Pain 13 (1999) 89–96.

[40] M.S. Matharu, A.S. Cohen, P.J. Goadsby, SUNCT syndrome responsive to intravenous lidocaine, Cephalalgia 24 (2004) 985–992, https://doi.org/10.1111/j.1468-2982.2004.00886.x.

[41] R. Gil-Gouveia, P.J. Goadsby, Neuropsychiatric side-effects of lidocaine: examples from the treatment of headache and a review, Cephalalgia 29 (2009) 496–508, https://doi.org/10.1111/j.1468-2982.2008.01800.x.

[42] C.S. Sung, S.H. Lin, K.H. Chan, W.K. Chang, L.H. Chow, T.Y. Lee, Effect of oral clonidine premedication on perioperative hemodynamic response and postoperative analgesic

requirement for patients undergoing laparoscopic cholecystectomy, Acta Anaesthesiol. Sin. 38 (2000) 23–29.
[43] C. Sun-Edelstein, A. Mauskop, Role of magnesium in the pathogenesis and treatment of migraine, Expert Rev. Neurother. 9 (2009) 369–379, https://doi.org/10.1586/14737175.9.3.369.
[44] M.D. Ferrari, P.J. Goadsby, K.I. Roon, R.B. Lipton, Triptans (serotonin, 5-HT1B/1D agonists) in migraine: detailed results and methods of a meta-analysis of 53 trials, Cephalalgia 22 (2002) 633–658, https://doi.org/10.1046/j.1468-2982.2002.00404.x.
[45] S.D. Silberstein, S.H. Kori, Dihydroergotamine: a review of formulation approaches for the acute treatment of migraine, CNS Drugs 27 (2013) 385–394, https://doi.org/10.1007/s40263-013-0061-2.
[46] A.J. Nagy, S. Gandhi, R. Bhola, P.J. Goadsby, Intravenous dihydroergotamine for inpatient management of refractory primary headaches, Neurology 77 (2011) 1827–1832, https://doi.org/10.1212/WNL.0b013e3182377dbb.
[47] T. Pringsheim, D. Howse, In-patient treatment of chronic daily headache using dihydroergotamine: a long-term follow-up study, Can. J. Neurol. Sci. 25 (1998) 146–150, https://doi.org/10.1017/s031716710003376x.

Index

'*Note:* Page numbers followed by "f" indicate figures and "t" indicate tables.'

N

O

P

R

S

Printed in the United States
by Baker & Taylor Publisher Services